Large-Scale Mammalian Cell Culture

Academic Press Rapid Manuscript Reproduction

These papers were presented as part of a Symposium on Large Scale Mammalian Cell Culture by the American Chemical Society Division of Microbial and Biochemical Technology at the 188th American Chemical Society National Meeting, Philadelphia, Pennsylvania, August 27, 1984.

Large-Scale Mammalian Cell Culture

Edited by

Joseph Feder and William R. Tolbert

Monsanto Company
St. Louis, Missouri

1985

ACADEMIC PRESS, INC.

Harcourt Brace Jovanovich, Publishers

Orlando San Diego New York Austin
London Montreal Sydney Tokyo Toronto

ACADEMIC PRESS, INC.
Orlando, Florida 32887

United Kingdom Edition published by
ACADEMIC PRESS INC. (LONDON) LTD.
24–28 Oval Road, London NW1 7DX

LIBRARY OF CONGRESS CATALOGING-IN-PUBLICATION DATA
Main entry under title:

Large-scale mammalian cell culture.

Papers presented as part of a symposium sponsored by the American Chemical Society Division of Microbial and Biochemical Technology at the 188th American Chemical Society National Meeting, Philadelphia, Pa., Aug. 27, 1984.

Includes index.

1. Cell culture—Congresses. 2. Biological products—Congresses. 3. Mammals—Cytology—Congresses. I. Feder, Joseph. II. Tolbert, William R. III. American Chemical Society. Division of Microbial and Biochemical Technology.

QH585.L36 1985 599'.00724 85-48071
ISBN 0-12-250430-5 (alk. paper)
ISBN 0-12-250431-3 (paperback)

PRINTED IN THE UNITED STATES OF AMERICA

85 86 87 88 9 8 7 6 5 4 3 2 1

Contents

Contributors

Numbers in parenthesis indicate the pages on which the authors' contributions begin.

G. D. Ball (87), *Wellcome Biotechnology Ltd, Beckenham, Kent, United Kingdom*

J. R. Birch (1), *Celltech Ltd, Slough, Berkshire, United Kingdom*

Ernest Bognar (39), *Research and Development Laboratory, K C Biological, Lenexa, Kansas*

R. Boraston (1), *Celltech Ltd, Slough, Berkshire, United Kingdom*

Peter C. Brown (59), *Bio-Response, Inc., Hayward, California*

Deborah H. Burke (127), *Verax Corporation, Hanover, New Hampshire*

P. Calabresi (125), *Roger Williams General Hospital, Providence, Rhode Island*

Maureen A. C. Costello (59), *Bio-Response, Inc., Hayward, California*

B. Creswick (125), *Roger Williams General Hospital, Providence, Rhode Island*

Robert C. Dean, Jr. (127), *Verax Corporation, Hanover, New Hampshire*

K. H. Fantes (87), *Wellcome Biotechnology Ltd, Beckenham, Kent, United Kingdom*

Joseph Feder (97), *Monsanto Company, St. Louis, Missouri*

N. B. Finter (87), *Wellcome Biotechnology Ltd, Beckenham, Kent, United Kingdom*

J. Hopkinson (125), *Amicon Corporation, Danvers, Massachusetts*

M. D. Johnston (87), *Wellcome Biotechnology Ltd, Beckenham, Kent, United Kingdom*

Subhash B. Karkare (127), *Verax Corporation, Hanover, New Hampshire*

K. Lambert (1), *Celltech Ltd, Slough, Berkshire, United Kingdom*

James L. Lewis (59), *Bio-Response, Inc., Hayward, California*

C. Lewis, Jr. (97), *Monsanto Company, St. Louis, Missouri*

Bjorn K. Lydersen (39), *Research and Development Laboratory, K C Biological, Lenexa, Kansas*

Lee A. Noll (39), *Research and Development Laboratory, K C Biological, Lenexa, Kansas*

Robert Oakley (59), *Bio-Response, Inc., Hayward, California*

Michael Patterson (39), *Research and Development Laboratory, K C Biological, Lenexa, Kansas*

John C. Petricciani (79), *Food and Drug Administration, Bethesda, Maryland*

A. W. Phillips (87), *Wellcome Biotechnology Ltd, Beckenham, Kent, United Kingdom*

Philip G. Phillips (127), *Verax Corporation, Hanover, New Hampshire*
Gordon G. Pugh (39), *Research and Development Laboratory, K C Biological, Lenexa, Kansas*
James Putnam (39), *Research and Development Laboratory, K C Biological, Lenexa, Kansas*
Randall G. Rupp (19), *Damon Biotech, Inc., Needham Heights, Massachusetts*
P. W. Thompson (1), *Celltech Ltd, Slough, Berkshire, United Kingdom*
William R. Tolbert (97), *Monsanto Company, St. Louis, Missouri*
R. S. Tutunjian (125), *Amicon Corporation, Danvers, Massachusetts*
P. J. White (97), *Monsanto Company, St. Louis, Missouri*
M. C. Wiemann (125), *Roger Williams General Hospital, Providence, Rhode Island*

Preface

The past few years have witnessed a rapid development of large-scale mammalian cell culture technology for the production of biologically important molecules. This has been motivated by the advent of such therapeutic candidates as interferon, urokinase, and monoclonal antibodies and by the progress in recombinant DNA technology. Genetically engineered animal cells that express glycosylated proteins, secreted in high yield in native conformation, have opened new opportunities for large-scale cell culture. This book examines this new technology, its potential for commercial application, and the regulatory concerns posed by its use for the production of human therapeutics.

THE LARGE SCALE CULTIVATION OF HYBRIDOMA CELLS PRODUCING MONOCLONAL ANTIBODIES

J. R. Birch
P. W. Thompson
K. Lambert
R. Boraston

Celltech Ltd
Slough, Berkshire
U.K.

INTRODUCTION

In 1975 Kohler and Milstein developed the hybridoma technique which allowed for the first time the production of monoclonal antibodies recognizing specific antigens of choice. The exquisite specificity of monoclonal antibodies combined with the potential for producing them in unlimited quantities has lead to their widespread application in many areas of biological research. They are also beginning to find commercial applications especially in the field of diagnostic assays, and increasing interest is being shown in their potential therapeutic uses and in their application to affinity purification systems. It has become apparent that some of these applications will require very large amounts of antibody, in some instances many kilograms. Table I gives an idea of the amounts of antibody that might be required. In the case of diagnostic applications, although the amount of antibody used in each test may be trivial, tens or hundreds of grams per year may be required if the test is widely used. Indeed the first large scale use of monoclonal antibodies was for ABO blood typing (Voak & Lennox 1983). For *in vivo* imaging if one assumes that a few hundred micrograms are used per patient (see for example Mach et al. 1981), then for each 10,000 patients one will require several grams of

TABLE I. Large Scale Antibody Requirements

Application	Range of requirements per annum (g)
Diagnostic assays	1 - 200
In vivo imaging	1 - 100
Immunotherapy	1,000 - 100,000
Immunopurification	1 - 1,000

antibody. If the requirement is for *in vivo* therapy then it is likely that least several hundred milligrams might be used per patient (see for example Miller et al. 1982). Hence for each 10,000 patients one will need several kilograms of antibody. Interestingly there is already a kilogram scale application of therapeutic monoclonal antibody in the veterinary area. Molecular Genetics Inc. have developed a monoclonal antibody-based therapy for Scours disease in newborn calves (Sadowski, 1982). In the case of immunopurification applications are still at a relatively small scale, but as antibodies become available in larger quantities and more cheaply we are likely to see an increasing interest in larger scale applications using perhaps kilograms of antibody (Hill et al. 1984).

In summary there are already many antibodies required in quantities from grams to hundreds of grams. In the future we will see requirements for kilograms of antibody and it is generally recognized that this will necessitate a new approach to production.

A. Production Methods for Monoclonal Antibodies

Laboratory scale production of monoclonal antibodies is generally carried out by growing hybridoma cells either in small scale cell culture systems such as flasks and roller bottles or in rodents as ascites tumors. Production in animals becomes less attractive as the scale of production increases. If we assume that on average one can produce 50 milligrams of monoclonal antibody from a mouse then it requires 20,000 mice to produce one kilogram of antibody.

Apart from the practical problems of handling very large numbers of animals there are other disadvantage of ascites route compared with cell culture and these are summarized in Table II.

Having opted for the cell culture route we are then faced with a plethora of possible production methods. The small scale methods which have traditionally been used, namely flasks and roller bottles, are not appropriate for producing large quantities above say one gram. Essentially two approaches have been taken towards large scale production. In the first case hybridoma cells are immobilized or entrapped and then perfused with a culture medium. Techniques have been described for entrapping cells in semipermeable hollow fibre bundles (Wiemann et al. 1983) within microcapsules (Geyer et al. 1984) or within agarose microbeads (Nilsson et al. 1983). To date there has been relatively little published information concerning these techniques and it is still too early to predict how generally applicable they will be or how easily they could be scaled up. In the second approach cells are in free homogeneous suspension in the nutrient medium.

TABLE II. Advantages of Cell Culture Over Ascites

Low levels or absence of extraneous antibody
Reproducible and consistent process
Large quantities can be produced at low cost
Reduced risk of contamination by adventitious agents of rodents
Production of human antibodies (difficult in rodents)

B. Homogenous Suspension Culture Systems

Our approach to large scale production of monoclonal antibodies has been to use homogeneous suspension culture in deep tank fermenters. Before describing this system in detail it will be useful to go through the criteria which we considered in opting for cell suspension as opposed to an immobilized cell system. These criteria are summarized in

Table III. Workers in the field of animal cell culture will be well aware of the hazards of microbial contamination. We therefore need a system that can be effectively sterilized prior to use and from which microorganisms can be excluded by appropriate barriers during prolonged operating periods which may extend from one or two weeks for a batch culture up to several months for a continuous culture process. Efficient mixing is important because we want to achieve uniform dispersion of the cells in the culture vessel and also to ensure that any additions we make to the culture (for instance acid and alkali for pH control) are rapidly and evenly distributed throughout the culture. At the same time the mixing has to be compatible with the fragile nature of the cells we are growing. Mixing will also influence the mass transfer characteristics of the reactor and possibly the most important factor here is oxygen supply. At atmospheric pressure oxygen is sparingly soluble in dilute aqueous solutions (approximately 7 mg/l for water saturated with atmospheric air). One million hybridoma cells would utilize this amount of oxygen in approximately one hour (Boraston et al. 1984). Hence we need to ensure effective transfer of oxygen to the culture if we are to avoid oxygen limitation of growth. Whatever system we choose has to be capable of being scaled up since we want to benefit from the economics of scale. A tenfold increase in reactor size might only increase capital expenditure by two to threefold and labor input will be significantly less compared with running many smaller vessels. A homogeneous system in which mixing and mass transfer characteristics

TABLE III. Criteria for Selecting Cell Culture Systems

Aseptic operation
Mixing compatible with shear sensitive cells
Adequate mass transfer characteristics, especially in relation to oxygen solution rate.
Facility for scale-up
Ease of process control and automation
Compatibility with upstream and downstream processing
Availability of ready-made process equipment

are understood facilitates prediction of process behavior at different scales. Such a system also offers the best option for direct monitoring and control of process parameters. Finally wherever possible we try to use standard equipment in order to minimize engineering development work.

Whilst the production of monoclonal antibodies in deep tank fermenters has received relatively little attention, conventional stirred fermenters up to 8,000 litres capacity have been used routinely for the production of foot-and-mouth disease virus and of interferon (for discussion see Phillips et al. in this volume).

C. Airlift Reactors

One type of fermenter which we have studied in detail and which satisfies the criteria described above is the airlift reactor. The growth of animal cells in this type of reactor was first described by Katinger et al. (1979) who considered it to have particularly appropriate characteristics for shear sensitive cells. The principle of the airlift is illustrated in Figure 1 and is described in more detail in a recent review (Smart, 1984). Gas mixtures introduced into the culture inside the base of the draught tube establish circulation of the culture because gas holdup causes a density differential between the contents of the draught tube and of the outer zone of the vessel. By varying the composition of this gas mixture the dissolved oxygen tension and the pH of the culture can be controlled. One significant advantage of the airlift system is that it obviates the need for moving parts and their mechanical seals which are a feature of turbine agitated reactors and a potential route for contamination. The typical airlift reactor is also capable of oxygen transfer rates which meet the measured oxygen demand of hybridoma cells (Boraston et al. 1984) Often there is a tendency to ignore the oxygen demand of animal cells in culture. Whilst their oxygen demand is indeed low compared with microbial cells we have found hybridoma cells to be obligately aerobic and some cell culture systems would not support cell growth at high cell population densities. For further discussion of the oxygen requirements of animal cells in culture refer to Fleischaker & Sinskey (1981) and Glacken et al. (1983).

We have worked with airlift vessels of 5, 30 and 100 litres working volume and have recently installed a 1000 litre fermenter. Twenty-two different cell lines of mouse, rat and human origin have been grown in these fermenters, all producing monoclonal antibodies. The total output in our first year of production from these fermenters exceeds 450 grams of purified or concentrated antibody. Scale-up has been found to be predictable in that growth kinetics and antibody production are similar at 5 and 100 litre scales. Thus we expect our 1000 litre vessel to have a capacity of at least 5 kilograms per annum.

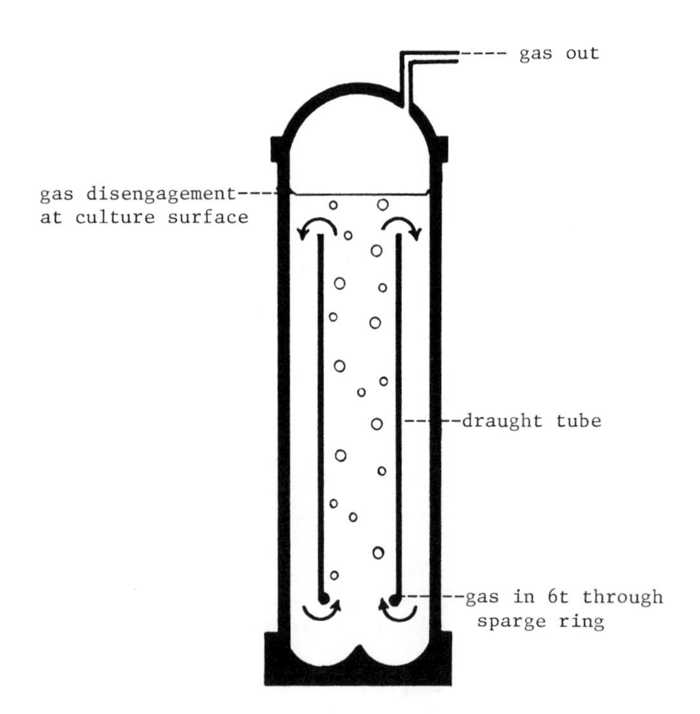

PRINCIPLE OF AIRLIFT FERMENT

Fig. 1. Gas introduced within the base of the draught tube causes a circulation of culture fluid upwards through the draught tube and down through the annular space between the draught tube and the vessel wall. Dissolved oxygen tension and pH can be controlled by varying the composition of the sparged gas.

By careful process optimization we have been able to increase the antibody yield for many hybridomas by four to five fold compared with static flasks or roller bottles. We find, like others, that antibody production varies greatly from one cell line to another. Our experience and the published literature (for instance Goding 1980) indicate that rodent cell lines typically produce between 10 and 100 milligrams of antibody per litre of culture fluid in static flasks. In our fermenters we see a range from about 40 to 500 milligrams per litre and the average yield from cell lines grown to date is 102 milligrams per litre.

Understanding the basis of the inherent variation in antibody synthesis between cell lines and devising methods for generating more productive cell lines is something we recognize as a high priority.

Figure 2 shows growth and antibody production by a rat hybridoma cell line, YOK 5/19, which produces an IgG antibody which is used for the assay and purification of interferon a (Hawkins et al. 1983). This result was obtained in a 30 litre airlift vessel (LH Engineering Ltd.). Dissolved oxygen tension, pH and temperature were all monitored and automatically controlled. After a brief lag phase the cells grew exponentially to a population density of 2×10^6 cells/ml and then entered a decline phase. Synthesis of antibody occurred during the growth phase and interestingly, continued during the decline phase. Indeed approximately 50 per cent of the antibody was produced after the maximum cell population density had been realized suggesting that antibody synthesis is not growth rate-dependent. The antibody was harvested immediately after the final time point shown on this graph. We know from other experiments that antibody synthesis does not continue beyond this point.

Most of our work has been carried out with rodent cell lines but we have also successfully grown human antibody producing cells in airlift culture. Human monoclonal antibodies are likely to be come increasingly important especially for therapeutic applications.

D. Continuous Culture

The batch culture of YOK 5/19 hybridoma cells illustrated in Figure 2 follows a classical profile for batch culture of most microorganisms, i.e. a brief lag phase followed by exponential cell growth and finally decline phase. Clearly under batch culture conditions the environment is constantly changing - nutrients become depleted and metabolites and cell

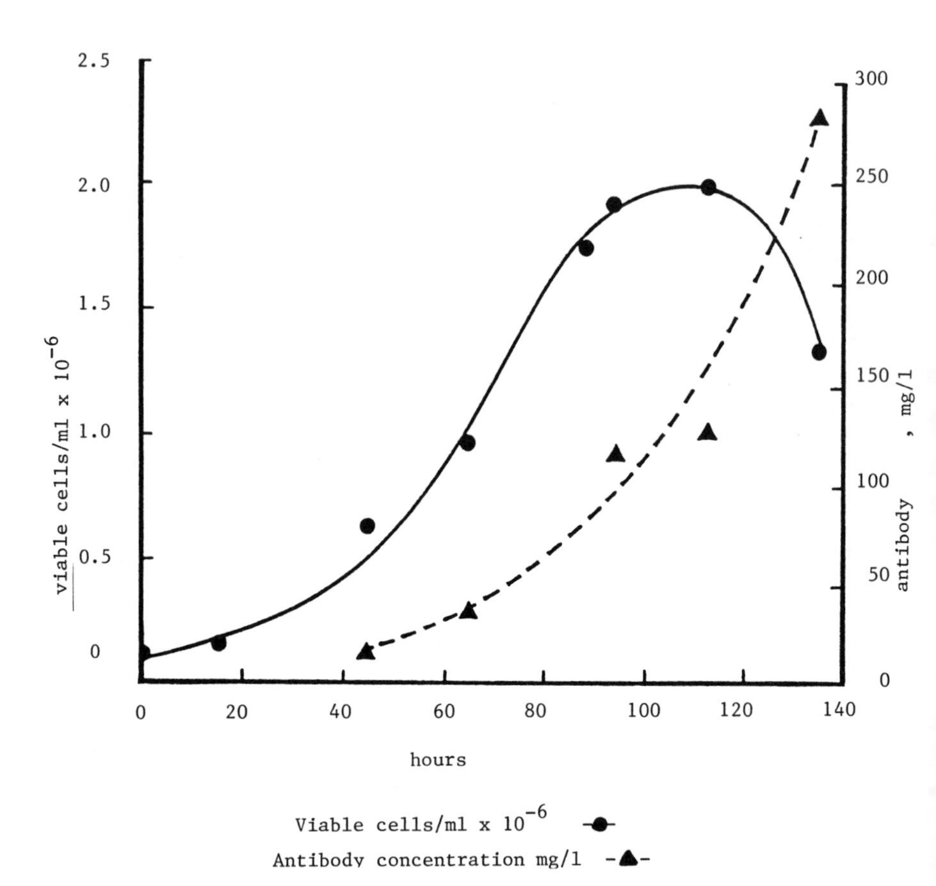

FIG. 2. Batch Culture of a Rat Hybridoma Cell Line, Yok 5/19, in an Airlift Fermenter.

products accumulate - and there are concomitant changes in a cell physiology. It is relatively difficult to analyse such a constantly changinng system and equally difficult to exert any influence on the physiology of the cells that might benefit the synthesis of antibody.

For this reason we have examined the utility of continuous chemostat culture, (Pirt, 1975; Tovey, 1980). Chemostat culture differs from batch culture in that a continuous feed rate. The culture will generally be started batch-wise and the medium feed initiated during the exponential phase of growth. An effluent of culture containing cells and cell products is harvested at the same rate. The result is that the cells achieve steady state conditions in which a state of constant exponential growth is maintained. Their specific growth rate can be controlled by varying the rate of medium feed and the cell population density can be deliberately dictated by medium design such that a single chosen nutrient or metabolite is at a growth limiting concentration. Providing the dilution rate (medium feed rate per unit culture volume) does not exceed the maximum growth rate of the cells then a physiological steady state is established in which any measured parameter (e.g. cell population density and viability, concentrations of nutrients and cell products) should be constant with respect to time.

We have used the chemostat to study cell growth and antibody synthesis for NB1 cells, a mouse hybridoma cell line which synthesizes an IgM antibody against human red blood cell group B antigen and is used as a blood typing reagent (Sachs & Lennox, 1981). A typical steady state for NB1 cells growing in oxygen limited chemostat culture is shown in Figure 3 and it can be seen that the measured parameters remain constant. This constant environment offers the facility to make accurate measurements of metabolite concentrations, metabolic quotients and product synthesis rates and further to investigate the effects of deliberately altering the environmental conditions.

The effects of varying growth rate under a variety of nutrient limitations upon antibody synthesis by NB1 cells is illustrated in Figure 4. Under glutamine limitation the specific rate of antibody synthesis is slightly depressed at low growth rates while under oxygen (Boraston et al. 1983) or glucose limitation this trend is reversed. The overall picture indicates that antibody synthesis is not growth rate dependent and this would substantiate the observation that in batch culture antibody synthesis continues during the decline phase of the culture.

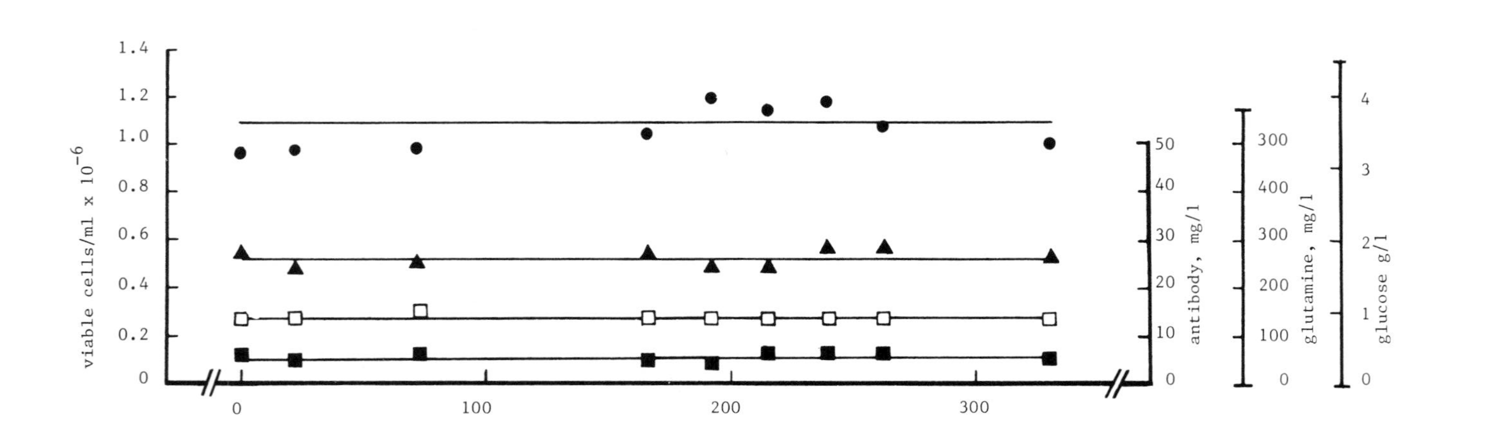

FIG. 3. A Typical Steady State for the Mouse Hybridoma Cell Line, NB1, Growing in Continuous (Chemostat Culture). Oxygen was the growth limiting nutrient and the specific growth rate was 0.021 h^{-1}. Concentrations of glucose and glutamine in the medium feed were 4.5 g/l and 584 mg/l respectively. Viable cells/ml x 10^{-6} -●- ; antibody concentration mg/l -▲-; glucose g/l -□-; glutamine mg/l -■- .

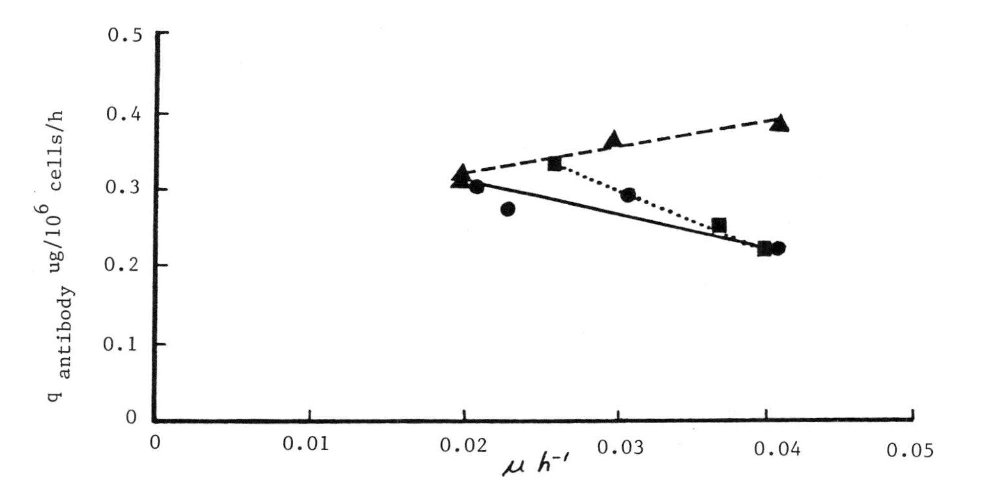

FIG. 4. Effect of Growth Rate (μ) on Specific Rate of Antibody Synthesis ($q_{antibody}$) for the Mouse Hybridoma Cell Line, NB1, Growing in Continuous (Chemostat) Culture. Growth limiting nutrients were glutamine (- - ▲ - -), glucose (... ■ ...) or oxygen (--●--).

For large scale antibody production continuous culture offers a number of advantages over batch culture and while our analytical studies have been performed at the 1-5 litre scale we are currently operating a 30 litre pilot scale continuous fermenter. Once established a continuous culture can be maintained at high cell density and with consequent high product output for many months, an inherently more productive and efficient system than one based on a batch process. The constant environment and the control of growth rate and nutrient limitations that can be attained in the chemostat permit deliberate manipulation of cell physiology so that there is far greater scope for optimization of biomass and product formation than can be achieved in batch culture. Fazekas de St. Groth (1983) has used continuous culture to grow ten different hybridoma cell lines. In this case cultures were run as turbidostats or cytostats so that the cells grew at or about their maximum growth rate and the medium feed rate was adjusted to maintain a constant cell population density. (Table IV)

One prerequisite for a successful continuous production system is that cell lines should be stable in terms of antibody synthesis. One would expect that chemostat culture might exert a selective pressure against cells which synthesize a product such as an antibody which is presumably nonessential and does not contribute to generation of new biomass. In the case of hybridoma cells antibody synthesis could be considered an unnecessary metabolic burden and one might expect nonproducing variant cells to appear in the culture and to come to predominance. We have taken NB1 cells from three separate chemostat cultures, recloned them by limiting dilution and then tested the clones for loss of antibody synthesis. Although a small proportion of nonproducing cells had been detected in the initial cell population there was no demonstrable increase in the proportion of nonproducers after between sixty and ninety generations in the chemostat even at low growth rates when the selective pressure would be most stringent (Table V).

TABLE IV. Advantages of Continuous (Chemostat) Culture

Constant environment.
Cells can be grown under a variety of nutrient limitations.
Control of growth rate.
More efficient use of fermenter capacity than batch systems.
Fewer manipulations.

TABLE V. Stability of Antibody Synthesis for the Mouse Hybridoma Cell Line, NB1, Growing in Continuous (Chemostat) Culture

Experiment	Nutrient Limitations	Hours	Generations	Clones Tested	Clones +VE	% +VE
Chemostat 1	Glutamine * Oxygen * Glucose *	1647	92	20	20	100
Chemostat 2	Glutamine * Oxygen *	1647	92	20	20	100
Chemostat 3	Glutamine	1647	60	24	23	96
Control	---	---	--	33	32	97

* indicates separate nutrient limitations applied to successive steady state.

E. Factors Affecting Antibody Purification

When considering the criteria for selecting a cell culture system we stated that the system should be compatible with downstream processing. Small scale purification of monoclonal antibodies has been well described in the literature and there are relatively few problems in handling small volumes of cell culture supernatant. However, novel problems do occur when handling large (100-1000 litres) batches of culture supernatant containing antibody in relatively dilute solution.

Our approach here has been to apply an initial rapid concentration step to the clarified culture supernatant. Using tangential flow ultrafiltration we can obtain at least a tenfold concentration with virtually quantitative recovery of antibody within less than one hour of harvesting from the fermenter. With the antibody now concentrated to between one and five grams per litre combinations of precipitation, ion exchange chromatography, gel filtration and affinity chromatography can be used to obtain antibody at the desired grade of purity. Using these techniques, greater than 95 per cent purity can routinely be achieved.

Most of the major contaminating proteins are derived from the serum which is used to supplement the growth medium so that purification of monoclonal antibodies can be facilitated by use of serum-free media. Elimination of serum by a variety of alternative medium supplements has been described (Chang et al. 1980; Cleveland et al. 1983; Fazekas de St. Groth, 1983; Kawamoto et al. 1983; Kovar & Franke, 1983; Murakami et al. 1983) and these supplements are summarized in Table VI. Serum-free media are now commercially available although they are generally expensive compared to conventional media and furthermore do not necessarily support cell growth in agitated culture systems. We have recently developed a serum-free medium which supports the growth of hybridoma cells in either static or agitated culture. Apart from the obvious advantages of a medium that is of defined and consistent composition, elimination of serum also reduces the chances for ingress of adventitious microorganisms, a particularly important consideration in the production of therapeutic reagents.

TABLE VI. Serum Substitutes for Hybridoma Cells

Albumin
Insulin
Transferrin
Hydrocortisone
Oleic acid (sometimes complexed to albumin)
Cholesterol (sometimes as low density lipoprotein)
Ethanolamine
Ascorbic acid

SUMMARY

We have developed a large scale production process for monoclonal antibodies based on deep tank fermentation. This allows us to produce large quantities of antibody economically and in a form which can be conveniently purified. We believe that there is still considerable scope for improving the productivity of the process both by continuing process optimization and by genetic manipulation of the producing cells. For very large scale applications it may well be that the ultimate route for production will be in microorganisms using recombinant DNA techniques. In this respect it is interesting to note that functional antibody has already been produced from bacterial cells (Boss et al. 1984).

REFERENCES

1. Boraston, R., Thompson, P.W., Garland, S. and Birch, J.R. (1984). Develop. Biol. Stand. 55, 103-111.
2. Boraston, R., Garland, S. and Birch, J.R. (1983) J. Chem. Tech. Biotech. 33b, 200.
3. Boss, M.A., Kenten, J.H., Wood, C.R. and Emtage, J.S. (1984). Nucleic. Acids Res. 12, 3791-3804.
4. Chang. T.H., Steplewski, Z. and Koprowski, H. (1980). J. Immunol. Methods. 39, 369-375.
5. Cleveland, W.L., Wood, I. and Erlanger, B.F. (1983). J. Immunol. Methods. 56, 221-234.
6. Fazekas de St. Groth, S. (1983). J. Immunol. Methods 57, 121-136.

7. Fleischaker, R.J. and Sinskey, A.J. (1981). Eur. J. Appl. Microbiol. Biotechnol. 12, 193-197.
8. Geyer, D.S., Collins, A.J., Koch, G.A. and Rupp, R.G. (1984). In vitro 20, No. 3, part 11, abstract 104.
9. Glacken, M.W., Fleischaker, R.J. and Sinskey, A.J. 1983. Trends in Biotechnology 1, 102-108.
10. Goding, J.W. (1980) J. Immunol. Methods 39, 285-308.
11. Hawkins, R.E., Spragg, J.S. and Secher, D.S. (1983). Antiviral Research, The Biology of the Interferon System. Elsevier Biomedical. Abstrsct 1, No. 2, 9.
12. Hill, C.R., Birch, J.R. and Benton, C. (1984), in Bioactive Microbial Products III - Downstream Processing. Special publication of the Society for General Microbiology. Academic Press, London and New York. In press.
13 Katinger, H.W.D., Sheirer, W. and Kromer, E. (1979). Ger. Chem. Eng. 2, 31-38.
14. Kawamoto, T., Sato, J.D., Le, A., McClure, D.B. and Sato, G.H. (1983). Anal. Biochem. 130, 445-453.
15. Kohler, G. and Milstein, C. (1975). Nature (London) 256, 495.
16. Kovar, J. and Franek, F. (1984). Immunology Letters 7, 339-345.
17. Mach. J-P., Buchegger, F., Forni, M., Ritschard, J., Berche, C., Lumbroso, J-D., Schreyer, M., Giradet, C., Accola, R. and Carrel, S. (1981). Immunology Today, December issue, 239-249.
18. Miller, R.A., Maloney, D.G., Warnke, R. and Levy, R. (1982). The New England Journal of Medicine 306, 517-522.
19. Murakami, H., Edamoto, T., Shinohara, K. and Omura, H. (1983). Agric. Biol. Chem. 47, 1835-1840.
20. Pirt, S.J. (1975). Principles of Microbe and Cell Cultivation, Blackwell Scientific Publications.
21. Nilsson, K., Scheirer, W., Merten, O.W., Ostberg, L., Liehl, E., Katinger, H.W.D. and Mosbach, K. (1983). Nature 302, 629-630.
22. Sacks, S.H. and Lennox, E.S. (1981). Vox Sanguinis 40, 99-104.
23. Sadowski, P.L. (1982). European Patent Application EP 0 077 734 A2.
24. Smart, N.J. (1984). Laboratory Practice. July issue.
25. Tovey, M.J. (1980). Advances in Cancer Research 33, 1-37.
26. Voak, D. and Lennox, E.S. (1983). Biotest Bulletin 4, 281-290.
27. Wiemann, M.C., Ball, E.D., Fanger, M.W., Dexter, D.L., McIntyre, O.R., Bernier, Jr., G. and Calabresi, P. (1983). Clin. Res. 31, 511.

DISCUSSION OF THE PAPER

DR. S. B. KARKARE (Verax Corporation, Hanover, N.H.): Can you give some kind of comparative figures for serum-free medium and normal medium?

DR. J. R. BIRCH: I do not have specific data which I can provide to you at present, except that in general terms the productivity appears to be very similar.

DR. S. B. KARKARE (Verax Corporation, Hanover, N.H.): Do you use antibiotics in the medium?

DR. J. R. BIRCH: Depending on the application, we generally do, although in commissioning all of our fermenters, we always run them for prolonged periods with antibiotic-free medium as a test of the system.

DR. W. S. HU (University of Minneapolis, MN): In your continuous culture, did you find strain degeneration a problem? Did you assay what fraction of cells are producers and what fraction are nonproducers?

DR. J. R. BIRCH: Yes, as I indicated in Table V, in the starting population there was, by limiting dilution, 33 clones examined. Thirty-two were producers and one was a non-producer clone. Essentially, by cloning after large numbers of generations in chemostat culture we picked up in one case one non-producing clone. There was no evidence that the number of non-producing clones had increased. It was a very rare event.

DR. W. S. HU (University of Minneapolis, MN): You showed one curve where the specific productivity was plotted versus dilution rate. The range of dilution rates was quite narrow, only a few-fold difference. Do you know if you had a wider range of dilution rate, would you still see the same kind of trend or not?

DR. J. R. BIRCH: Well, what we have found for the nutrient limitations that we've looked at, the cells showed rather interesting characteristics and that is that the cells are only capable of growing over a fairly narrow range of specific growth rate. There seems to be a minimum specific growth rate at which these cells are capable of growing. If you work at extremely low dilution rates, then the viability of the cultures decreases dramatically.

DR. G. BELFORT (Rensselaer Polytechnic Institute, Troy, NY): Could you make some comments on the optimum cell concentration at which to operate for particular hybridoma lines that you have examined? The reason I ask is that it appears from some of the work that you referenced that it's not always optimum to operate at the highest cell concentration. Can you comment on that?

DR. J. R. BIRCH: Obviously the cell is the catalyst that is making antibodies. So all other things being equal, the more cells you've got the more antibody you ought to be making. In those situations where antibody synthesis actually goes down, I guess what you're saying is that some other environmental factors change such as dissolved oxygen or whatever to cause decrease in synthesis. I can't specifically comment on it because we've not seen in our systems anything that specifically occurs when we have very high cell densities which decreases antibody synthesis. It's not something we have observed. I guess that it is something that relates to the oxygen supply or some other aspect of the environment which as yet has not been analyzed.

USE OF CELLULAR MICROENCAPSULATION IN LARGE-SCALE PRODUCTION OF MONOCLONAL ANTIBODIES

Randall G. Rupp

Damon Biotech, Inc.
Needham Heights, MA

The purpose of this paper is to discuss the scale-up of hybridoma cells for antibody production in cell culture. There are obvious situations in which cell culture is necessary for monoclonal antibody production, e.g., in the culture of human-human hybridomas for which there is no appropriate animal model for ascites production. Even if there were, production from ascites fluid is fraught with technical and regulatory problems. On the other hand, large-scale culture of hybridoma cells has inherent technical problems similar to those of any cell culture system. Most cultured cells, for instance, require serum supplemented growth medium. This serum is a major source of contaminating proteins in the purification of any secreted cellular product. In addition, cultured cells usually grow to relatively low densities - 10^7 cells/ml of medium is high for cultured animal cells. Since the amount of product is directly proportional to the number of cells in culture, the volumes of harvested media necessary to purify adequate amounts of products become cumbersome.

Table I lists some important factors to be addressed in the scale-up to production of biologicals. Each of these must be considered as early as the research stage. A process that meets some, but not all, of these criteria has limited commercial potential.

I wish to discuss our approach to these problems in large-scale cell culture and specifically our use of microencapsulation in these endeavors.

Table I. Considerations in Scale-Up to Production

- Type and quality of desired product
- Type and flexibility of technology
- Space requirements and limitations
- Labor intensity
- Regulatory requirements

Figure 1 shows a typical hybridoma growth curve. It demonstrates the exponential growth of cells followed by a sudden decrease after day 5 or 6 with little or no plateau phase. This phenomenon must be overcome for the scale-up of hybridoma cells to production levels.

Figure 2 demonstrates that this "drop-dead" phenomenon can be eliminated by culturing the cells in a stirred tank reactor in which the requisite oxygen is provided through the head space and the medium changed continuously by pumping fresh medium in and spent medium out at the same rates. Both elimination of oxygen (at day 19) and discontinuance of the medium feed (at day 23) caused the total cell number and viability to drop.

Clearly, by properly balancing medium exchange and cell number and maintaining sufficient oxygen, a cytostatic culture can be maintained. Cytostatic cultures of hybridoma cells and their associated production of monoclonal antibodies has been beautifully detailed by Fazekas de St. Groth (1). As demonstrated by the relatively constant cell number from day 30 to 39 a cytostatic culture has been achieved in this culture. The spent medium that was collected contained 35 µg of antibody per ml and can be processed to produce relatively pure monoclonal antibodies.

There are three major problems however in using cytostatic cultures for large-scale production of monoclonal antibodies:

1. The size of the required culture vessels is relatively large.
2. The volumes of medium to be processed are large and therefore necessitate a concentration step.
3. Contaminating serum proteins and serum associated antibodies co-purify with the monoclonal antibody.

Various approaches have been taken to minimize these problems. The most fruitful has been to increase the cell density thereby increasing the concentration of the antibody and decreasing the volumes of spent medium to be processed. Cell densities have been increased by developing cell culture reactors that permit more efficient cellular utilization of nutrients. The most common reactor modifications and process

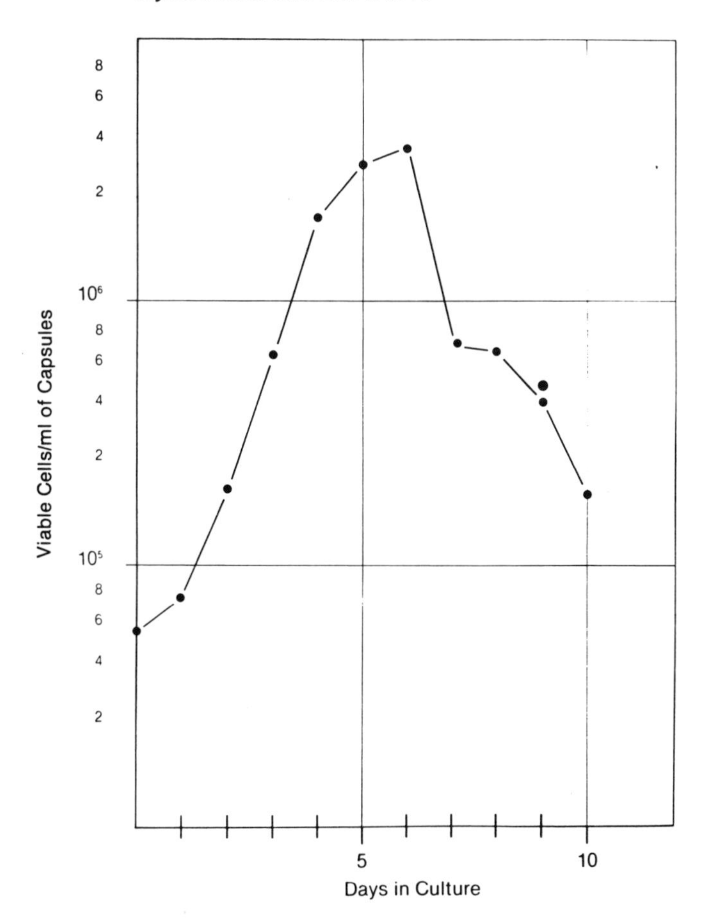

Figure 1. Hybridoma growth curve. Hybridoma cells were seeded at 4.5×10^4 cells/ml of medium in 3-liter spinner flasks and incubated in a humidified incubator at 37°C with a 95% air/5% CO_2 atmosphere. At specified times, 5 ml samples of the suspension were taken and cells counted on a hemacytometer.

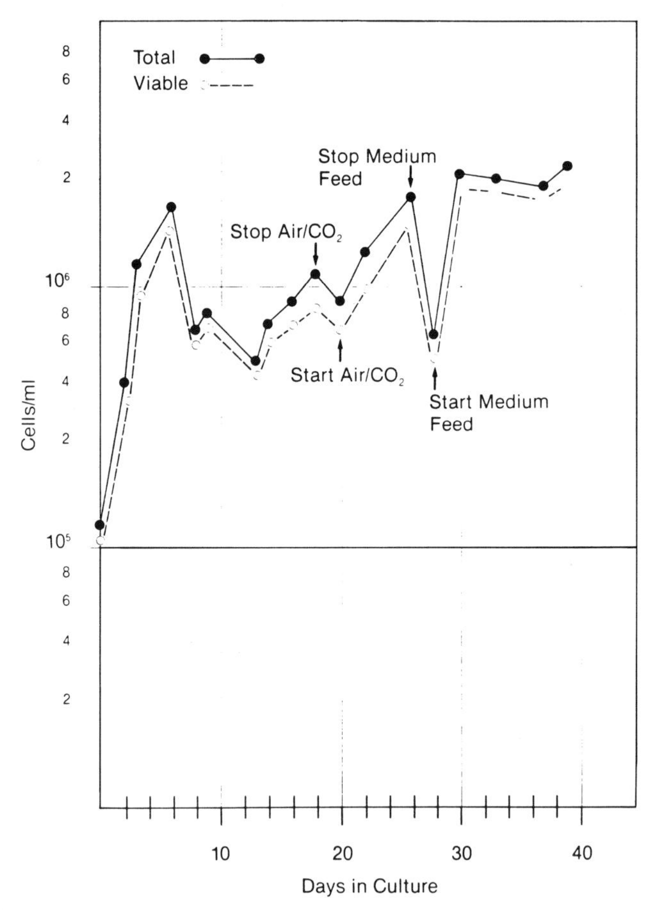

Figure 2. Cytostatic culture of hybridoma cells. Hybridoma cells were seeded at 1.2 x 10^{-5} cells/ml in 15 liters of medium in a 40-liter stirred-tank reactor with 95% air/5% CO_2 provided through the head space. At specified intervals, 5 ml samples were collected, treated with 0.2% trypan blue and counted on a hemocytometer. At day 4, fresh medium was pumped into the vessel at 4 ml/min and spent medium and cells removed at the same rate. On day 19 of culture, the air/CO_2 was discontinued and started again on day 20. On day 20 the rate of medium exchange was increased to 6 ml/min. On day 23, the medium pumps were stopped and restarted on day 24. From day 30 to the end of the culture at day 39, the rate of medium exchange was 10 ml/min.

changes in non-anchorage dependent cells are (a) the addition of spin-filters to conventional stirred tank reactors,(2); (b) the use of hollow fibers as culture chambers (3); and (c) the use of cellular microencapsulation (4). These have been discussed and compared by Glacken *et al* (5) and Margaritis *et al*. (6).

This report discusses the use of microencapsulation as a way to circumvent and minimize some of these problems. Cellular microencapsulation was first developed and discussed by Lim and Sun (7) and has been used subsequently for many applications (8). Basically, the procedure involves placing living cells inside an alginate sphere coated with a semipermeable membrane which permits adequate diffusion of nutrients to the cells and cellular anti-metabolites away from the cells. Figure 3 summarizes the essential steps in the microencapsulation process

Microencapsulated cells are cultured in stirred-tank reactors at a microcapsule-to-culture volume ratio of 20-50%. The cell culture reactor has been specifically designed to encapsulate, culture and harvest the microencapsulated cells. Figure 4 is a cross sectional diagram of the reactor vessel. The head plate contains ports for sterile additions and venting of gases, for the placement of an oxygen probe, and for sterile sampling. The top drive has a programmable acceleration start-up feature to minimize damage induced by starting the stirrer in a settled bed of microcapsules or microcarriers. Because the cells are protected from physical shear inside the microcapsules, direct sparging of gases is permitted.

Medium is pumped continuously to the vessel or in batches, through the bottom center port and permitted to overflow into a stand-pipe located within a stainless steel screen mesh. The 500 micron diameter microcapsules are retained within the vessel. At the termination of the culture the medium is drained through a port located at the bottom of the vessel within the screen mesh. This port along with the bottom center port permits washing of the microcapsules with sterile solutions before harvest through the central port.

Higher cell densities are obtained (with less labor intensity) when the cultures are perfused continuously rather than batch-fed (see Figure 5). A comparison of the cell densities obtained in a continuously-fed microcapsule culture and in a conventional cytostat culture is seen in Figure 6. The microcapsule culture attains densities in excess of 108 cells/ml of microcapsules as compared to 10^6 cells per ml of medium in the cytostatic suspension culture.

Table II summarizes medium consumption and antibody

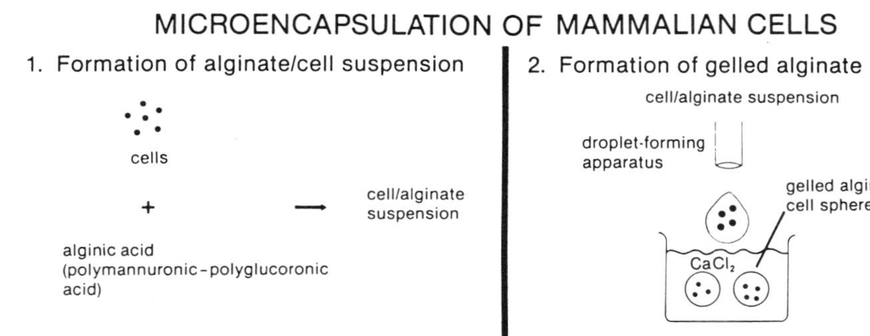

Figure 3. Schematic representation of the microencapsulation of hybridoma cells.

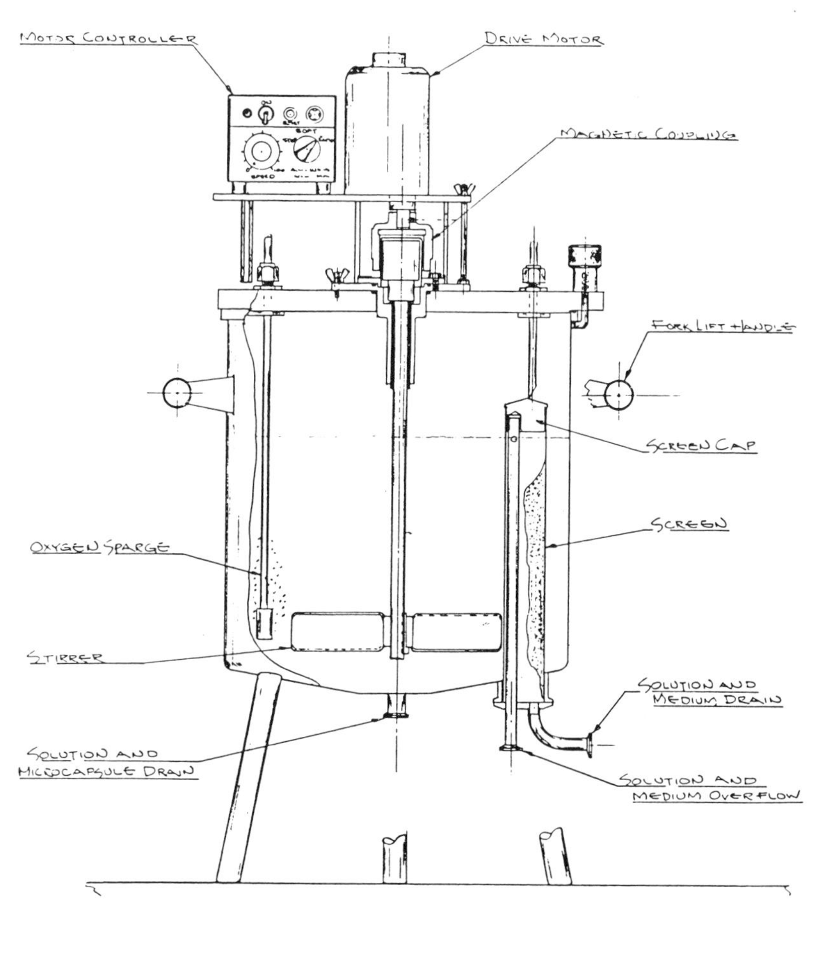

Figure 4. Cross-sectional diagram of the bioreactor vessel.

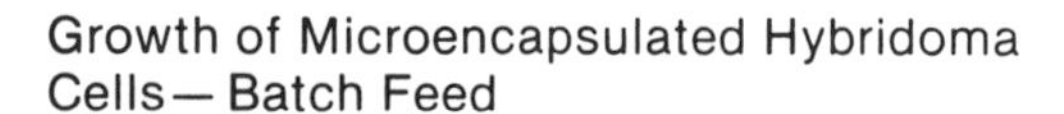

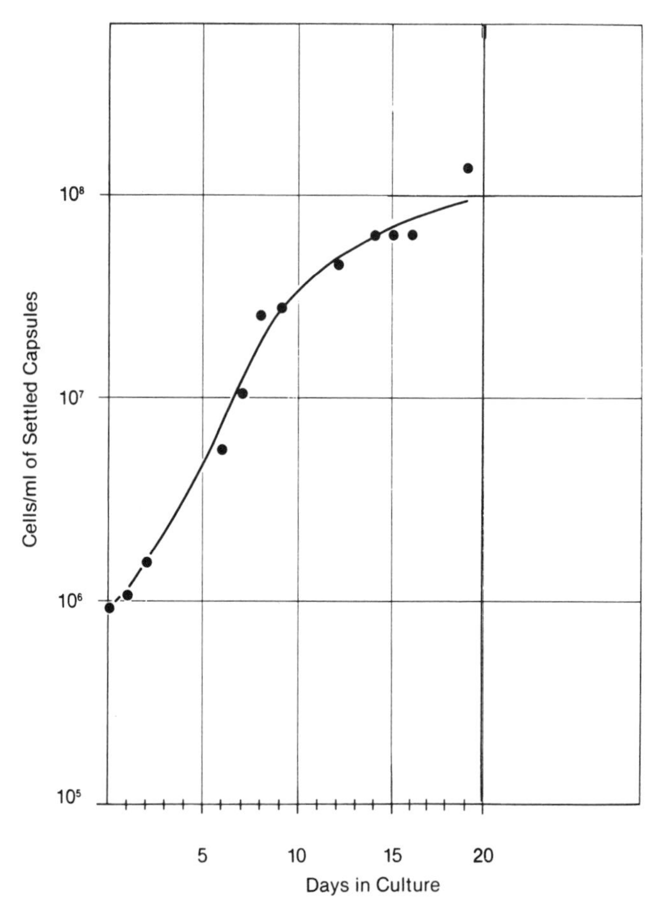

Figure 5a. Batch vs continuous feed of microencapsulated hybridoma cells. Hybridoma cells were microencapsulated to yield 4 liters of settled microcapsules. The batch was divided into two, 2-liter lots which were placed in 2 separate bioreactors. Each was incubated at 37°C with direct sparging of 95% air/5% CO_2. In one vessel (5a), 50% of the medium was removed and replaced with fresh medium every other day beginning on day 60 of culture. In the other vessel (5b) the medium was continuously exchanged at 4 ml/min from day 4-7, at 7 ml/min from day 8-12 and at 10 ml/min from day 13 to the end of the culture. Samples were taken and cells counted as described in Figure 2.

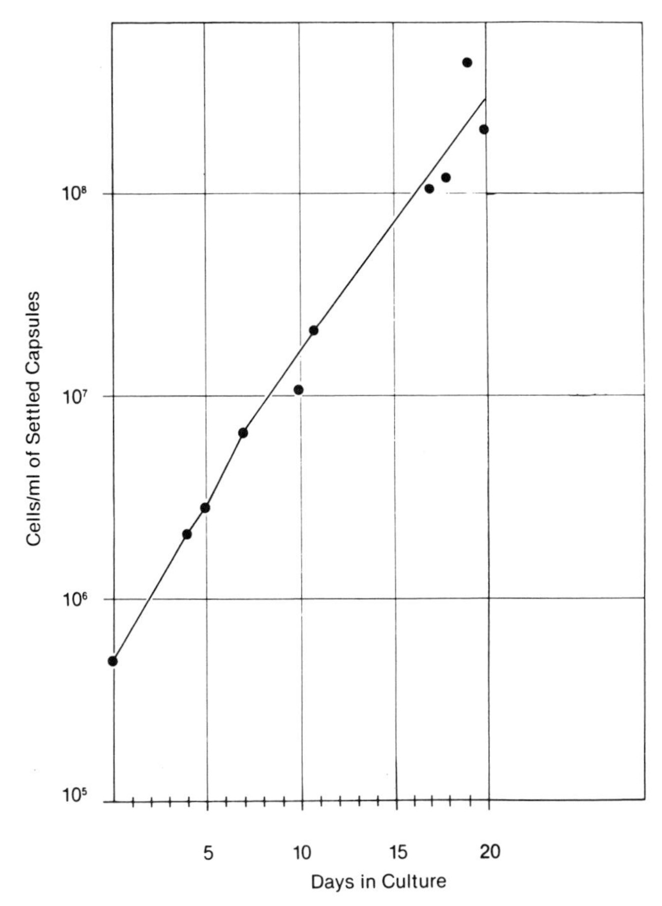

Figure 5b. Batch vs continuous feed of microencapsulated hybridoma cells.

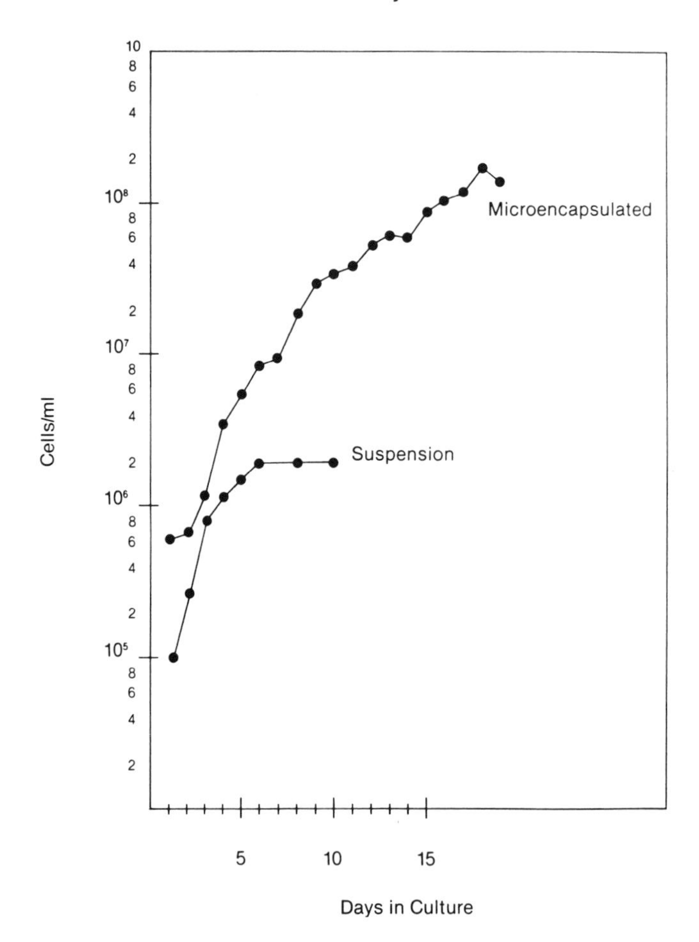

Figure 6. Comparison of hybridoma growth curves in suspension and microcapsule culture. Microencapsulated cells were cultured and cells counted as described in Table II.

Table II. Cell-line dependence of Medium Consumption and Antibody Production

Hybridoma	Volume of microcapsules (liter)	Cell Density (cells/ml of capsule	Total cells/ vessel	Medium Consumed (liters)	Antibody Yield (g/liter microcapsule	Antibody Yield (g)
Mouse/ mouse						
BL-1-LH	2.0	4.0×10^{7}	8.0×10^{10}	170	3.0	6.0
BL-3-OD	2.0	1.45×10^{6}	2.9×10^{9}	200	10.0	20.0
100-84	4.0	7.0×10^{7}	2.8×10^{11}	120	1.5	6.0
102-84	2.4	8.9×10^{11}	2.1×10^{11}	250	4.0	9.6
106-84	2.4	4.8×10^{7}	1.2×10^{11}	210	2.75	6.6
108-84	2.3	1.1×10^{8}	2.5×10^{11}	270	4.3	9.9
100-83	4.0	3.0×10^{8}	1.2×10^{12}	240	1.5	6.0

Legend to Table II. Hybridoma cells were cultured in a bioreactor with continuous exchange of medium and direct sparging of 95% air/5% CO_2. At the end of the culture period, the microcapsules were washed 5 times with pyrogen-free saline and opened by differential homogenization. Cell densities were determined on a 1 ml aliquot by counting the cells on a hemocytometer. Intact cells and capsular debris were removed from the remaining intracapsular material by centrifugation at 14,000 X g for 10 minutes. Aliquots of the clarified intracapsular material were assayed for antibody by nephelometric and enzyme-linked immunosorbent (ELISA) assays.

production data from 7 cell lines grown using this system. Clearly, the amount of medium consumed and antibody produced is cell line dependent.

The kinetics of intracapsular antibody accumulation is shown in Figure 7. The accumulation of antibody is proportional to the number of viable cells within the microcapsule. In this particular example the rate of antibody production remains constant at 12,000 molecules of antibody/cell/minute during this culture period as is shown in Figure 8. This clearly demonstrates a lack of feedback inhibition by the accumulated of antibody or anti-metabolites.

Intracapsular antibody purity shown as a percentage of total protein increases during culture to approximately 50% (Figure 9). During the culture period the non-antibody protein remains relatively constant at approximately 1 mg/ml of microcapsules. The increased purity with time is due to a preferential increase in antibody (data not shown).

The effect of the initial purity of antibody on its subsequent purification by ion-exchange chromatography can be seen in Figure 10. With starting purities in excess of 20%, a final purity of greater than 99% is obtained with one pass of the material over an ion-exchange column.

Affinity column purification has been and is now being used for the purification of many monoclonal antibodies. We have avoided the use of affinity purification for three main reasons:

1. The covalently bound antigen or antibody tends to leach into the desired "captured" product when it is stripped from the matrix. The amount of material carried into the product is difficult to ascertain, thereby making validation of this procedure for production of a human-injectable antibody difficult if not impossible.
2. Some applications of monoclonal antibodies require a patient-specific antibody. To minimize the possibility of cross contamination of patient antibodies, a different affinity column for each patient would be required. This would be extremely expensive.
3. Many monoclonal antibodies lose a significant amount of their antigen-binding capacity as a result of the conditions required to remove them from the affinity matrix. Table III compares monoclonal antibody purified by affinity and by ion-exchange chromatography. Antibody purified by affinity chromatography lost between 32% and 52% of its antigen binding capacity

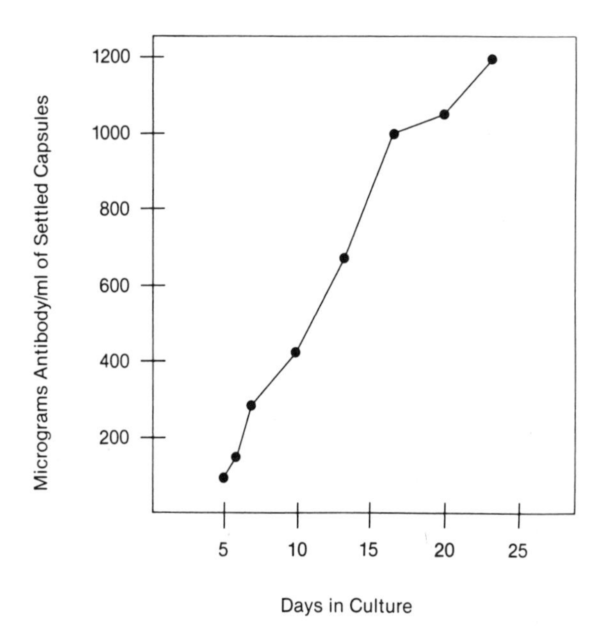

Figure 7. Kinetics of intracapsular antibody accumulation. Microencapsulated hybridoma cells were cultured as decribed in Table II. Microcapsule samples were taken as indicated and assayed for antibody by ELISA.

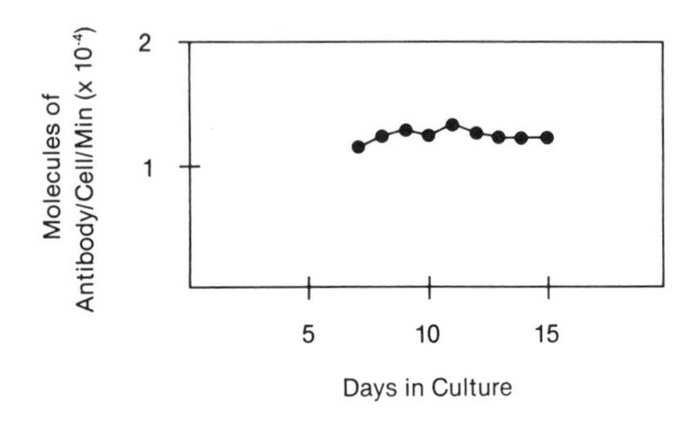

Figure 8. Rate of antibody production. Microencapsulated hybridoma cells were cultured and assayed for antibody and cell density as described in Table 2. The rate of antibody production is the difference of cell counts between 2 specified times divided by the rate of cellular antibody production was estimated by dividing the difference in intracapsular antibody concentration between 2 specified times by the difference in cell density during the same time. (Aliquots of the extracapsular culture medium were sampled and assayed for antibody and were determined to contain less than 5% of the total antibody in the culture, thereby, indicating there was not significant leakage of antibody from the microcapsules.

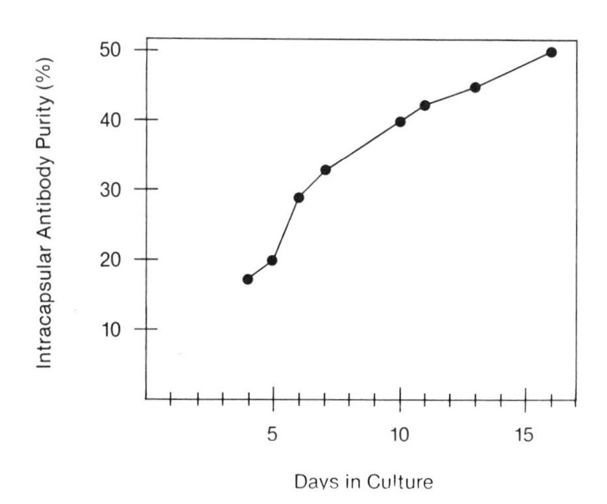

Figure 9. Intracapsular antibody purity as a function of culture time. Microencapsulated hybridoma cells were cultured and assayed for total cells and antibody as described in Table II. In addition, total intracapsular protein was determined by the method of Lowry et al (9). Antibody purity was expressed as the ratio of total intracapsular antibody to total intracapsular protein X 100%.

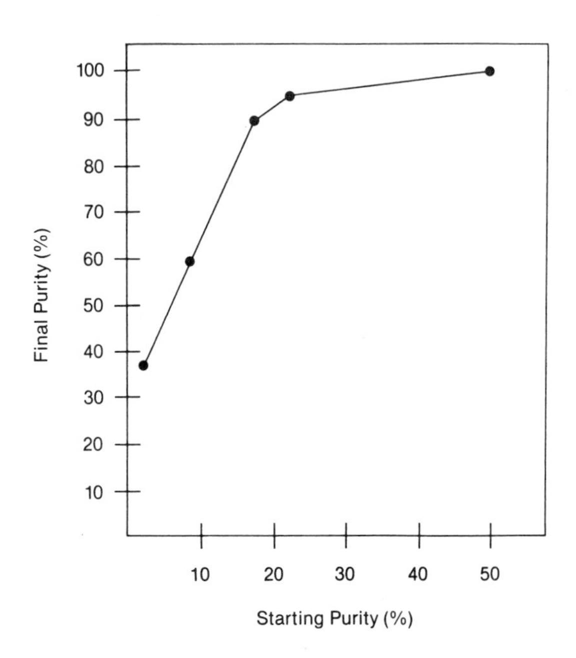

Figure 10. Effect of initial antibody purity on subsequent purification. Intracapsular materials with different antibody purities were concentrated separately by precipitation with 60% ammonium sulfate and collected by centrifugation at 15000 X g for 15 minutes. The ammonium sulfate pellets were dissolved and dialyzed exhaustively in a Tris/NaCl buffer. Samples of each were passed over DE-53 cellulose - ion exchange columns. The unbound material, containing the antibody, was collected and assayed for total protein and antibody concentration. Antibody purity was expressed as the mass ratio of antibody to total protein X 100%.

when removed from the matrix with glycine/HCl and acetate buffers, respectively. In addition, the respective antibody recoveries were only 33% and 50%. Antibody purified by ion exchange chromatography lost no detectable antigen binding capacity. The purity of the antibody using either method was in excess of 98%.

In summary, we have found that microencapsulation and culture of hybridoma cells for production of monoclonal antibodies addresses the points made in Table 1. It is flexible, labor and space efficient cost effective and permits production of monoclonal antibodies that are functional and highly purified. The entire process uses materials that have been well established for use in the production of therapeutic agents.

Table III. Antibody Activity Following Purification of Intracapsular Fluid By Affinity or Ion Exchange Chromatography

	Starting Purity	% Antibody Activity	% Recovery	Final Purity
Affinity Chromatography	33%			
Acetate		32%	93	>98%
Glycine/HCl		52%	33	>98%
Ion Exchange	33%	100%	95	>98%

Legend to Table III. Intracapsular antibody was purified by passing it over a DEAE, Tris-acryl column or over an affinity column to which the antigen to the antibody was covalently coupled. The "captured" antibody was eluted from the affinity matrix with either 0.1 M acetic acid (pH 2.8) or 0.1 M glycine/HCl (pH 2.5) and was collected into tubes containing Tris and therefore minimize denaturation buffer (pH 7.5) to neutralize the antibody as a result of the low pH. The activity of purified antibody, prepared by either ion-exchange or affinity column, was determined by passing purified antibody over the antigen-affinity column and eluting with 0.1 M acetic acid. Total antibody protein recovered from the column was determined by the method of Lowry, et al.(9). The activity was estimated as total antibody protein (bound to the affinity column) divided by total antibody placed on the column x 100%.

REFERENCES

1. Fazekias de St. Groth, S. (1983). Automated Production of Monoclonal Antibodies in a Cytostat, J. of Immun. Methods, 57: 121-136.
2. Himmelfarb, P., Thayer, P., and Martin, H.E. (1969). Spin Filter Culture: The Propagation of Mammalian Cells in Suspension, Science, 164: 555-557.
3. Calabresi, P., McCarthy, K., Dexter, P., Cummins, F., and Rotman, B. (1981). Monoclonal Antibody Projection in Artificial Capillary Cultures. Proc. Amer. Assoc. Cancer Res. 22: 362-365.
4. Rupp, R., Posillico, E., and Gilbride, K. (1984). Manuscript in preparation.
5. Glacken, M., Fleischaker, R., and Sinskey, A. (1983). Mammalian Cell Culture: Engineering Principles and Scale-Up, Trends in Biotechnology, 1: 102-108.
6. Margarites, A. and Wallace, Jr., (1984). Novel Bioreactor Systems and their Applications, Biotechnology 2: 447-453.
7. Lim, F. and Sun, A., (1980). Microencapsulated Islets as Bioarticiial endocrine Pancreas, Science, 210: 908-910.
8. Jarvis, A. and Grdina, T. (1983). Production of Biologicals from Microencapsulated Living Cells, Biotechniques 1, 22-27.
9. Lowry, O., Rosenbrough, N., Farr, A., and Randall, R. (1951). Protein Measurement with the Folin Phenol Reagent. J. Biol. Chem., 193: 263-275.

DISCUSSION OF THE PAPER

DR. R. LYLE (Southwest Research Institute, San Antonio, Texas): Dr. Rupp, since your cell density seems to reach a maximum limit, does this mean that you do not have to worry about rupture of the microcapsules because of internal cell pressure?

DR. R. G. RUPP: That's true. As a matter of fact, we've grown bacteria in these microcapsules and the capsules are just teeming with bacteria. We don't see those breaking, even under high-power microscopes. We don't see anything.

DR. R. LYLE (Southwest Research Institute, San Antonio, Texas): Are there expansions of the microcapsules?

DR. R. G. RUPP: No. There is not. They remain constant throughout the culture.

DR. J. FIESHKO (Amgen Inc., Thousand Oaks, CA): I'm wondering about two things. First of all, you said that the overflow tube was fairly satisfactory, but have you considered using a load cell instead to control the volume, because the foaming can possibly cause a problem?

DR. R. S. RUPP: No. We were given certain parameters under which we had to work and since we were planning to produce a number of patient-specific antibodies, we needed to cram as many of these vessels into as small a space as possible. Load cells became somewhat prohibitive.

DR. J. FIESHKO (Amgen Inc., Thousand Oaks, CA): Have you looked at the viability of the cells inside the beads with time? It appears that you had a constant rate of production of antibody even though the cell density was increasing, suggesting that the specific productivity was decreasing. Do you think that is a viability related phenomena?

DR. R. G. RUPP: The actual calculation, I think, was done correctly. It was based on viable cells. But to answer your question, honestly, the cell viability is decreasing from about day 12. The problem which we are currently addressing is that the microcapsules are too big. Nevertheless we are getting good yields from microcapsules that are 500 microns. We are trying to make microcapsules that are 100-200 microns. I think what's happening is that the cell mass is so large that there's a limitation in diffusion to the center cells in the microcapsule. We have made smaller capsules and it has helped tremendously.

DR. A. W. PHILLIPS (Wellcome Biotechnology Ltd., Langley Court, Bedenham, Kent, England): I am interested in your comment about affinity purification of these antibodies. Were you in a position to distinguish between destruction of antibody on elution or just failure to elute when you got those low recoveries?

DR. R. G. RUPP: No, we were getting it all back but it was just not functional; this is not the case for all antibodies. We have done this experiment with 11 antibodies and seven of them reacted like this; four of them were perfectly fine, with acetate buffer. We've used isothiocyanate to strip it and we had the same results as with the sodium acetate.

DR. J. FEDER (Monsanto Company, St. Louis, MO): On one of your tables you showed, regarding one of the monoclonal productions, total production of about 6 grams. The density of the cells in the microcapsules was 10^8/ml. Obviously, the density of the cells in the total reactor volume is significantly less. You indicated a total medium utilization of about 170 liters, which means that the yield per liter of medium was about 35 mg/liter. Could you make some comments about the nature of that medium, concentration of serum, etc.?

DR. R. G. RUPP: The basal medium that we generally use is a modified William's essential medium. To that medium we add supplements and more amino acids. We found that that medium with normal concentration of amino acids allows for a certain cell density. That medium with twice the normal concentration of amino acids needed the same cell density, so the amino acids had no effect on cell density. However, with a twofold concentration of amino acids, the antibody production went from 200 micrograms/ml to about 1100 micrograms/ml, clearly indicating that amino acids are rate limiting. We don't know which one it is, how many if more than one, but it was a simple modification to get that higher yield.

DR. J. FEDER (Monsanto Company, St. Louis, MO): Does your medium contain serum?

DR. R. G. RUPP: Yes, there's 5% horse serum. We recently reduced that significantly though. With the continuous feed there's no need to put so much serum in the medium.

DR. A. J. PARCELLS (Upjohn, Kalamazoo, MI): You mentioned that you are using two different anion exchangers, trisacryl and Whatman. Do you have any preference of one over the other?

DR. R. G. RUPP: Yes, I like trisacryl. It doesn't compress when you strip it and it binds pyrogens if they're there.

DR. C. L. COONEY (Massachusetts Institute of Technology, Cambridge, MA): You mentioned in the preparation of beads that the algenic acid was retained inside the bead to give you a very viscous environment. How important is that high viscosity in getting good growth and production from the cells inside the microcapsule.

DR. R. G. RUPP: I don't think high viscosity is an important parameter, but to be honest, I don't know. We have never been able to eliminate enough of the alginate inside the microcapsule to do the experiment where we significantly reduced the alginate. The problem is alginate is a million molecular weight. If we put on a membrane to retain 80,000 molecular weight, we are obviously going to retain most of the alginate. So we're trying to address that problem now and we've been looking at it, but we haven't solved it yet.

THE USE OF A CERAMIC MATRIX IN A LARGE SCALE CELL CULTURE SYSTEM

Bjorn K. Lydersen
James Putnam
Ernest Bognar,
Michael Patterson
Gordon G. Pugh
Lee A. Noll

Research and Development Laboratory
K C Biological
Lenexa, Kansas

Two forms of ceramic matrix supporting the growth of both adherent and non-adherent animal cells to uniformly high density have been incorporated into an automated system for large scale cell culture. One form of the ceramic provides for attachment and growth of adherent cells and the complete harvesting of viable cells from the matrix. A different, more complex ceramic promotes immobilization of non-adherent cells as well as adherent cells, but harvesting of cells is difficult to perform. The system which has been developed utilizes a computer and two sets of sensors to continuously monitor and control pH and dissolved oxygen during the recirculation of medium through the ceramic culture chamber. This approach permits the maintenance of optimum medium conditions and the use of an effective method for quantitative monitoring through the continuous measurement of the oxygen consumption rate of the culture.

Three approaches have been successfully used to increase the scale of cell cultures; an increase in the amount of surface area provided by the ceramic results in proportional increases in cell yield, the use of the more complex ceramic results in greater yields, and multiple ceramics on a single system provide for proportional increases in scale.

Results demonstrating the utility of immobilization of non-adherent cells such as antibody synthesizing hybridomas are presented. Cultures can be maintained for long periods of time in low serum or serum-free media with either batch or continuous feeding. Antibody production rates are elevated with continuous feeding. A correlation between the oxygen consumption rate and rate of antibody synthesis has been observed, suggesting a novel approach to the maximization of productivity.

INTRODUCTION

Large-scale animal cell culture is used in the production of viral vaccines and many medically important proteins such as interferon, immunoglobulins and mammalian hormones and enzymes. Many approaches have been taken to promote efficient growth and productivity of the cells. Although a multiplicity of small cultures in flasks or roller bottles is still used for the production of many vaccines, several methods permitting single large cultures have been developed. For cells which can grow without adhesion to a surface, suspension cultures analogous to those used in fermentation of bacteria have been successful (1,2). Cells which require adhesion can also be cultured in stirred suspensions through the use of microcarriers as supports for cell attachment (3-5). Approaches utilizing microencapsulation of cells (6) or entrapment of cells in the walls of hollow fibers (7,8) have been developed, and these provide certain advantages for both adherent and non-adherent cells.

A new ceramic material has recently been developed which provides a very high degree of surface area per unit volume for the culture of adherent animal cells (9). This ceramic matrix has been shown to support growth of a wide variety of cells and to be effective in scale-up to at least $18.5m^2$ of surface area. Cultures on this ceramic are readily harvested. Through computer analysis of dissolved oxygen levels in the system, a quantitative method of culture monitoring is provided. It was found, however, that non-adherent cells were not immobilized to a high degree by this ceramic, limiting its breadth of application. In this report, results with a new form of ceramic matrix show that a high degree of immobilization of non-adherent cells can be obtained. Also, results on the scale-up of cultures on ceramic matrices are presented.

RESULTS

The two types of ceramic matrix used for cell culture are shown in Figure 1. Each is a cylinder with multiple square channels passing through it. The ceramic indicated as type A is useful for adherent cells and allows the complete harvesting of cultured cells. Due to differences in configuration, composition and texture, the ceramic indicated as type B provides for the immobilization of non-adherent cells. With this ceramic, however, only a portion of the cells can be harvested. The degree of cell growth on these ceramics is measured indirectly through analysis of the cellular oxygen consumption rate (OCR). Both types of ceramic have been incorporated into an automated system providing for control of pH, dissolved oxygen concentration and media flow rates as described in Figure 2. There are only minor differences in the methods and equipment used for the two types of ceramic.

Scale-up

Three approaches to scale-up have been investigated by varying the size of the ceramic, the nature of the ceramic, and the number of ceramics incorporated into a single system. Each approach has specific advantages, and through a combination of these approaches, systems suited to individual applications can be constructed.

It was previously shown (9) that increasing the dimensions of the ceramic matrix resulted in an increase in yield of Vero cells which was directly proportional to the increase in surface area. For each size of ceramic up to $18.5m^2$, the cell density increased from 2.3×10^4 cells/cm^2 on day zero to an average of 6.7×10^5 cells/cm^2 after 7-8 days of culture.

For industrial applications, two ceramics having $4.25m^2$ or $11m^2$ surface area were chosen for development. In four separate experiments with BHK cells, the cell yields and kinetics of oxygen consumption were compared for these ceramics as described in Figure 3. The larger ceramic was inoculated with three times as many cells and provided with three-fold more medium. The number of viable cells harvested from these ceramics was 3.2×10^{10} ($4.25m^2$) and 10.8×10^{10} ($11m^2$) cells after six days. The approximately

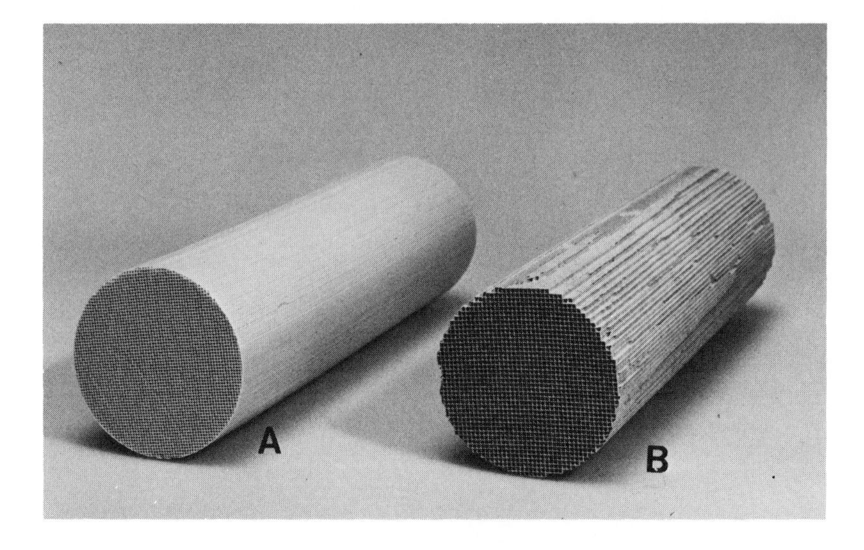

Figure 1. Ceramic matrices used for cell culture. Both types, indicated as A and B, are cylinders with uniform square channels passing through the length of the cylinder. Type A contains 68 channels per cm^2 of cross-sectional area, resulting in $32cm^2$ of surface area per cm^2 of volume. The type A ceramic shown has a radius of 4.1cm and length of 30cm, resulting in a total surface area of $4.25m^2$. The type B ceramic has 30 channels per cm^2 of cross-sectional area and varies in composition and texture from type A. The topographic features of the surface prevent a direct calculation of surface area available for cell attachment.

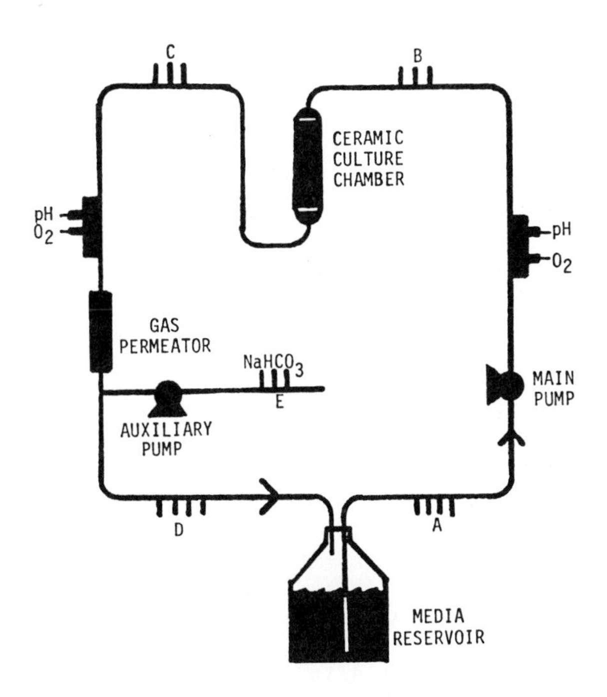

Figure 2. Schematic of system used for cell culture. The system provides continuous recirculation of medium through the encapsulated ceramic matrix, control and monitoring of pH and dissolved oxygen, and sterile access to the system for necessary operations such as cell inoculation, medium changes, and cell harvesting. Medium is recirculated (as shown by the arrows) with a peristaltic pump (Cole-Parmer) utilizing silicone tubing. Dissolved oxygen and pH are measured with probes from Instrumentation Laboratories and Ingold respectively. The gas permeators provide for equi-libration of the medium to controlled levels of CO_2 and O_2 without direct sparging of the medium by permitting exchange of gases through diffusion across the walls of silicone tubing. The access ports (labelled A-E) consist of silicone tubing with quick connect devices at the ends, covered by a sealable teflon bottle for maintenance of sterility. The control of pH and dissolved O_2 in this system is accomplished by an electronic comparison between the actual values (measured by the sensors prior to the inlet to the ceramic) and predetermined set-points. Deviations from the set points initiate changes in the relative percentages of the N_2, O_2 and CO_2 gas supplied to the permeators, or cause addition of $NaHCO_3$ to the medium.

three-fold greater number of cells harvested from the larger ceramic indicates scale-up was accomplished without losses in efficiency. The OCR patterns from the cultures (Figure 3) also indicate a roughly three-fold difference, confirming the results from the cell harvest and demonstrating the value of OCR in judging the relative size of the culture.

In addition to increasing the dimensions of the ceramic, scale-up can also be approached by varying the nature of the ceramic itself. In the experiment described in Figure 4, 10^9 MDCK cells were inoculated into either ceramic type A or B (both 8.2 x 30cm, DxL) and maintained with the same initial volume of medium. Although the apparent growth rates differed significantly when judged by the OCR's, the cultures were maintained for a sufficient length of time so that plateaus in the OCR developed in both cases. The plateau levels of the OCR did not increase significantly after the medium changes on the sixth day, suggesting the maximum number of cells had been attained on both ceramics. The magnitude of the OCR was roughly three-fold greater with ceramic B, suggesting that three times as many cells were present as on ceramic A. Since cells cannot be harvested from ceramic B, the observed increase in growth rate and cell number was not directly verified. However, experiments are being conducted to determine if productivity of the cells in terms of virus production and secretion of protein is in proportion to the observed differences in OCR. These results point out that the nature of the ceramic is a major factor in the relative capacity and performance of the ceramic.

The third method of scaling up within a single system is to utilize multiple ceramics. In an experiment to test this approach, five type A ceramics ($11m^2$) were maintained in parallel and the results compared to a system containing a single ceramic of $11m^2$. As indicated in Figure 5, BHK cells grew to a maximum OCR of 275 µmoles/min on the single ceramic. With an initial number of five times as many cells, the system with five ceramics supported growth to an OCR of 1400 µmoles/min, suggesting a direct correlation between the number of ceramics and capacity of the system. Cells were not efficiently harvested from the multiple system, preventing a direct comparison of cell numbers. Further testing of this approach to scale-up will focus on productivity in terms of virus or specific protein production.

Non-Adherent Cells

When cell lines such as SP2/0-Ag14, which have little tendency to adhere to surfaces, were inoculated in ceramic matrix

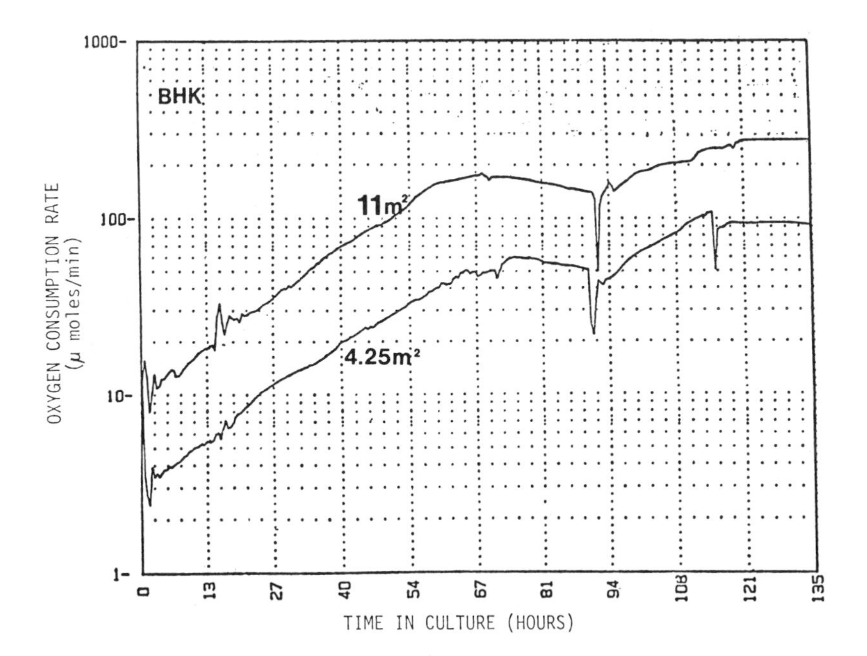

Figure 3. Oxygen consumption rates of BHK cultures. Ceramics (type A, Figure 1) having $4.25m^2$ and $11m^2$ surface area were inoculated with 1×10^9 and 3×10^9 cells respectively and incubated with DME medium supplemented with 8% FBS on the system described in Figure 2. The OCR was continuously calculated from the relationship: OCR (umoles/min) = $kF(\Delta 0^2)$, where k = 1.88 umoles/L, F = media flow rate (L/min), and $\Delta 0_2$ = difference in 0_2 probe readings in units of % of ambient ambient level of dissolved oxygen (Po_2 = 145mm Hg). The initial volume of medium was 10L ($4.25m^2$) and 30L ($11m^2$), with a replacement of 10L and 30L respectively after 88 hours. On the seventh day, the cells were harvested with 0.25% trypsin and enumerated by hemacytometer analysis. 3.1 x 10^{10} cells were recovered from the smaller ceramic (7.3 x 10^5 cells/cm^2) and 10.8 x 10^{10} cells from the larger ceramic (9.8 x 10^5 cells/cm^2). During the cultures, pH was regulated between 7.0 and 7.2 and dissolved 0_2 levels between 50-145mm Hg ($4.25m^2$) and 20-200mm Hg ($11m^2$).

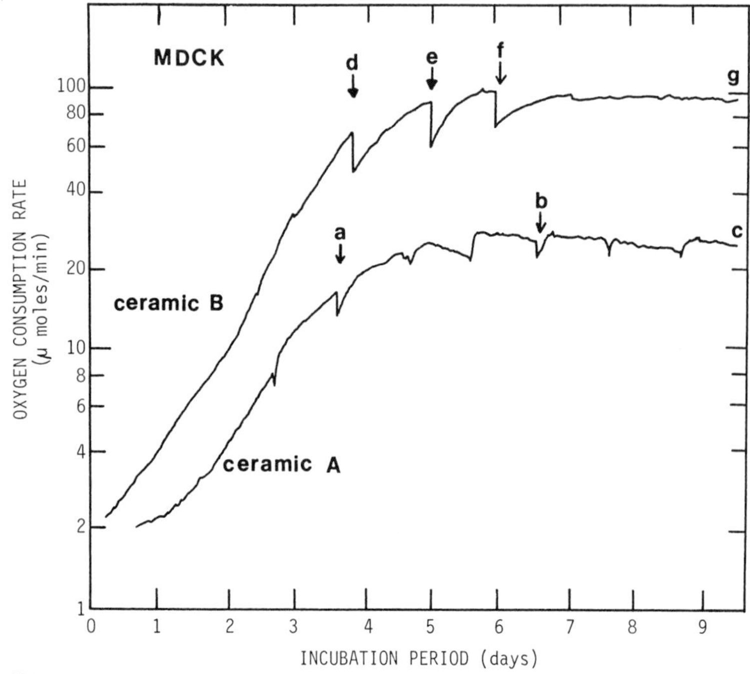

Figure 4. Oxygen consumption rates of MDCK cultures on two types of ceramic matrix. Cultures of MDCK cells in DME medium supplemented with 10% FBS were initiated by inoculating 10^9 cells into the two types of ceramic described in Figure 1. Each ceramic had dimensions of 8cm x 30cm (D x L); the number of channels per cm^2 were 68 (A) and 30 (B). The OCR was continuously calculated as described in Figure 3. The initial volume of medium was 11L for each culture. The arrows indicate the time at which 80% of the medium was removed and 11L of fresh, pre-warmed medium was added ("f" indicates the addition of 20L). Glucose concentrations in the medium at the times of exchange (measured on a Beckman Glucose Analyzer) were; a) 2770, b) 2650, c) 3160, d) 1100, e) 1200, f) 1400, and g) 1100mg/L (initial concentration was 4200mg/L). During the culture, pH was controlled at 7.2-7.3 and PO_2 was maintained in the range 100-150mm Hg (A) and 75-150mm Hg (B). 14.4 x 10^9 cells (96% viable) were harvested from ceramic A. Cells cannot be quantitatively harvested from ceramic B. In parallel to the culture on ceramic A, 850cm^2 roller bottles (Corning) were maintained with an initial cell number of 2 x 10^7 cells and the same ratio of medium to surface area. The cell number on these bottles after 10 days was 27 ± 2.4 x 107 (N = 3): the glucose concentrations were similar to those observed for the culture on ceramic A.

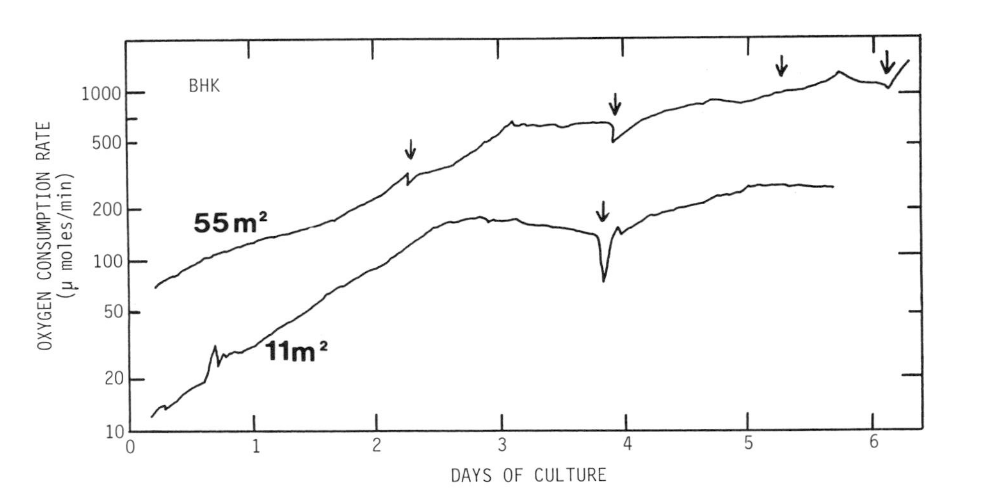

Figure 5. Oxygen consumption rates of BHK cells cultured on systems containing one or five culture chambers. BHK cultures were initiated on ceramics of $11m^2$ surface area (type A) with 3×10^9 cells per ceramic. Medium was recirculated through the ceramics with a peristaltic pump at rates of 0.2-1L/min per ceramic. A single pump was used for the system containing five ceramics in parallel. Oxygen consumption rates of the entire culture were continuously monitored, and are plotted in the figure. The arrows indicate times at which medium changes were made. A total of 60L DME with 8% FBS was used for the culture on the single $11m^2$ ceramic, and 185L was used for the system with multiple ceramics. Dissolved oxygen and pH levels were maintained at similar values in the two systems.

type A, an initially high percentage of immobilization was observed. In the experiment described in Figure 6, over 70% of SP2/0-Ag14 cells initially attached, but after 3 days no more than 20% of the culture was observed to be adherent to the ceramic. When ceramic B was tested with a mouse hybridoma (20-8-4S, ATTC HB 11) very similar in adhesive properties to SP2/0, quite different results were obtained, as shown in Figure 6, panel B. Initial attachment was greater than 95%, and the number of cells in suspension was approximately tenfold less than in a suspension culture seeded in parallel with the same total number of cells per unit volume of medium. Several parameters including antibody levels, amount of glucose consumed from the medium, oxygen consumption by the culture on ceramic, and staining of ceramics all confirmed that the great majority of the culture was immobilized.

Maintenance of this hybridoma line for an extended period of time on ceramic B was performed as described in Figure 7. After a period of rapid growth in medium with 10% FBS, the medium was changed to a serum-independent medium (KC2000™) containing lipids, serum albumin, transferrin and fetuin (total protein of 0.25mg/ml). When the OCR was observed to be decreasing, fresh batches of KC2000 were added (20-40L as described in the legend). In each case, rapid recovery of cellular oxygen metabolism was observed as judged by analysis of the OCR. Although the levels of OCR differed at the time fresh medium was added, five consecutive batches of fresh medium resulted in recoveries of the OCR to levels between 90-100 umoles/min. These results showed that long-term cultures could be performed with the ceramic providing for immobilization of the cells.

The number of cells in suspension was very low during this 43-day culture compared to the number in a parallel suspension culture. As shown in Figure 8, less than one-tenth as many cells were in suspension in the system containing the ceramic. Although cell concentration and viability were relatively high in the suspension culture during the first five days, the culture did not survive longer than a total of seven days at high density. In contrast, the culture could be maintained for at least seven weeks when immobilized on the ceramic.

The productivity of the cultures is compared in Figure 9. Although the initial concentrations of antibody in medium from the suspension culture and the system with the ceramic were similar, continued production was observed with the ceramic. The higher level of antibody observed in medium added at 600 hours was due to a lower volume supplied and an additional two day incubation period. In the final two batches of medium containing 1% FBS in place of KC2000, antibody levels

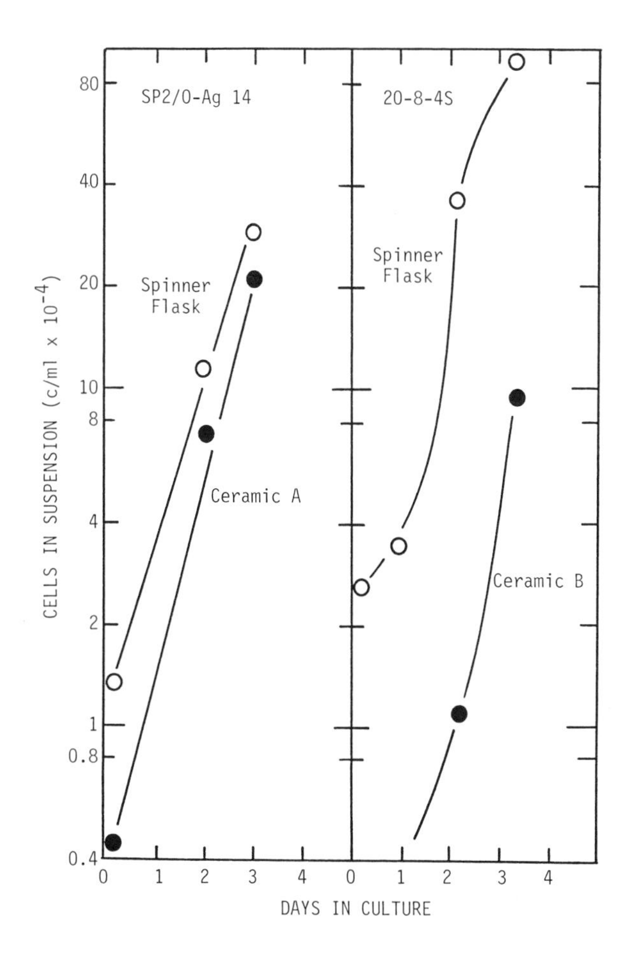

Figure 6. Number of cells in suspension in spinner culture compared to systems containing two types of ceramic. Panel A: SP2/0 (ATCC CRL 1581) cells were inoculated into a 0.5L spinner vessel (0) and a system containing the type A ceramic (0) at 1.3 x 10^4 cells/ml. Panel B: Mouse hybridoma cells 20-8-4S (ATCC HB 11), derived from a fusion between SP2/0 and C3H lymphocytes, were inoculated into a 0.5L spinner vessel (0) and the system containing the type B ceramic (0) at 2.5 x 10^4 cells/ml. In both experiments, the medium was RPMI-1640 supplemented with 10% FBS. Samples from the medium were analyzed with a hemacytometer for viable cell numbers.

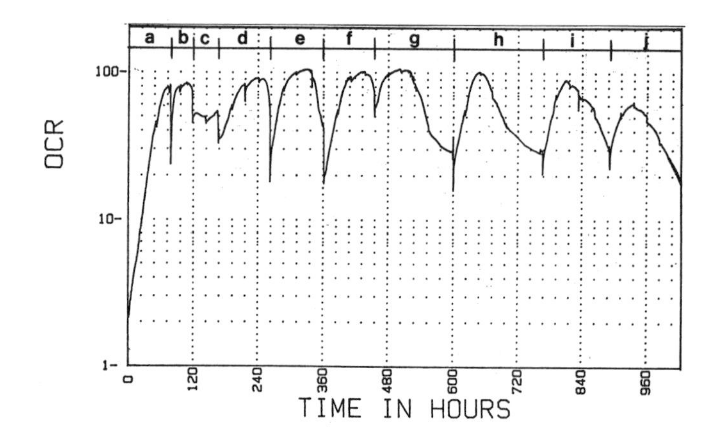

Figure 7. Oxygen consumption rate (OCR) of hybridoma cells immobilized on ceramic matrix. Hybridoma line 20-8-4S (ATCC HB 11) was inoculated into ceramic B (10^9 cells) and cultured for 3 days in 40L RPMI-160 medium supplemented with 10% FBS (a). Subsequently, the cells were maintained with the following media and volumes for the periods indicated by the lower case letters in the figure; b) 20L KC2000, c) 20L KC2000, d-g) 40L KC2000, h) 20L KC2000, i-j) 20L modified Iscove's DME, 1% FBS. KC2000 contains (Part A) a modified Iscove's DME (Part B) supplement of lipids, transferrin, serum albumin and fetuin (Part B). The OCR is used to monitor the metabolism of the culture. The correlation be-tween OCR and number of cells has not been determined for this cell line but data from cell lines such as BHK and Vero indicates 0.3-0.7 x 10^9 cells per unit of OCR (μmoles/min), suggesting roughly 5 x 10^{10} cells may be immobilized on the ceramic.

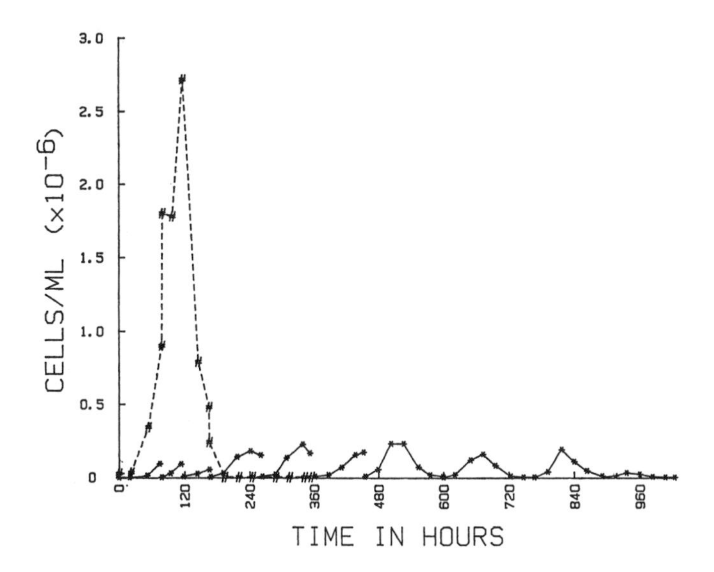

Figure 8. Hybridoma cells in suspension during culture in spinner flask and on the ceramic matrix. During the culture of cell line 20-8-4S on ceramic as described in Figure 7, samples of the medium were analyzed daily for number of viable cells (*). The suspension culture (#) was initiated with 2.5×10^4 cells/ml in 1L; after 3 days, the cells were centrifuged and resuspended in 0.5L of KC2000 to parallel the media change and volume reduction for the culture on ceramic as described in Figure 9.

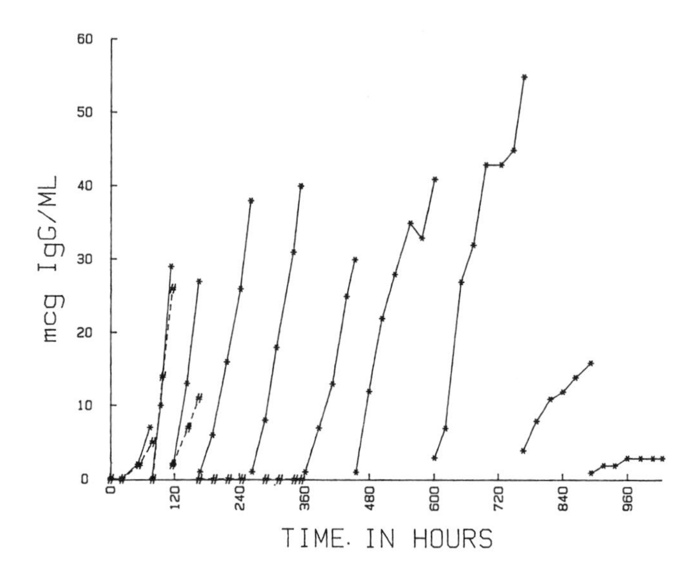

Figure 9. Concentration of antibody in medium during culture of hybridoma cell line immobilized on ceramic (*) or in suspension (#). During the cultures described in Figures 7 and 8, antibody concentration was determined with a quantitative ELISA assay, using purified mouse IgG as a standard.

were very low. In subsequent experiments however, 1% FBS supported similar production rates to that observed with KC2000.

When the amount of antibody synthesized per day was determined, it was observed that in many cases a strong correlation existed between production rate and OCR. This is readily observed between 180-480 hours. Although the maximum production rate in this period was about 500mg IgG/day, the average was significantly less. For improvements in the production rate per day, the data indicated that maintenance of OCR at the maximum level might be productive (Figure 10)

This was attempted in subsequent experiments by maintaining a continuous feeding and removal of medium at the same average rate per day as used in the batch feeding approach. As shown in Figure 11, this approach led to a culture in which the OCR was relatively stable. During a 14 day period the OCR gradually increased 30% in comparison to the batch-fed culture in which an over 50% change occurred in each of four cycles. Antibody production rates were significantly higher in the continuously fed culture. As shown in Figure 12 (panel A), the cumulative amount of antibody produced was approximately twice that in the batch-fed culture. The amount of medium supplied to the culture was equivalent in both experiments, as shown in Figure 12B, indicating the productivity of the system was doubled without an increase in the amount of medium utilized. These initial results with continuous feed and harvest suggest one way in which improved process control can result in improved productivity of large scale animal cell cultures. It is expected that manipulation and better control of other parameters will also lead to improvements in productivity.

DISCUSSION

The incorporation of ceramic matrices into the automated system described in this report provides a broadly applicable method for large scale production of biologicals from cell culture. The immobilization of both adherent and non-adherent cells to the ceramic provides distinct advantages in comparison to the maintenance of cells in a homogeneous suspension culture. With the rigid ceramic matrix, shear forces on cells are reduced since there is no need to maintain a suspension of cells or beads. This is particularly important for some cell types, especially when low protein media is used. With the cells limited to a discrete portion of the fluid system,

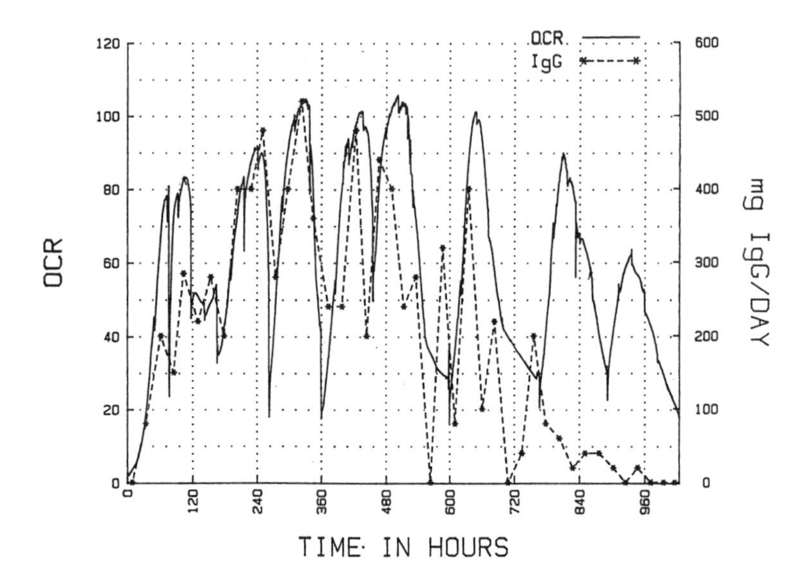

Figure 10. Antibody production rate of hybridoma cell line immobilized on ceramic matrix. The amount of antibody synthesized per day by the culture described in Figure 7 is plotted in comparison to the OCR.

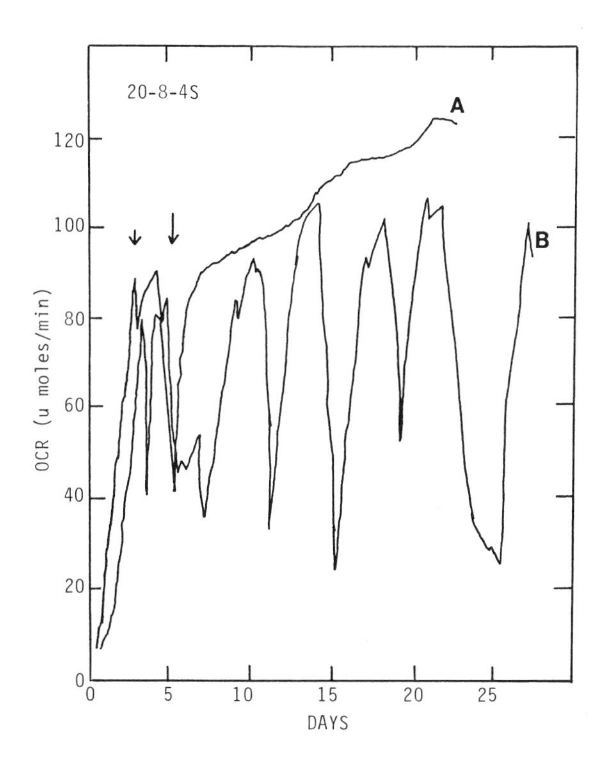

Figure 11. OCR's of immobilized hybridoma cultures maintained with batch or continuous feed. 10^9 hybridoma cells (20-8-4S) were used to initiate cultures on ceramic type B. After three days growth in 40L RPMI-1640 supplemented with 10% FBS, the medium was changed to: A) 20L modified Iscove's medium with 1% FBS (small arrow) and then maintained with continuous feed and harvest at a rate of 12L/day after the fifth day (large arrow), or B) 20L KC2000 and maintained with batch feeding as detailed in Figure 9. With this cell line, the rate of synthesis of IgG was found to be similar in KC2000 and medium with serum.

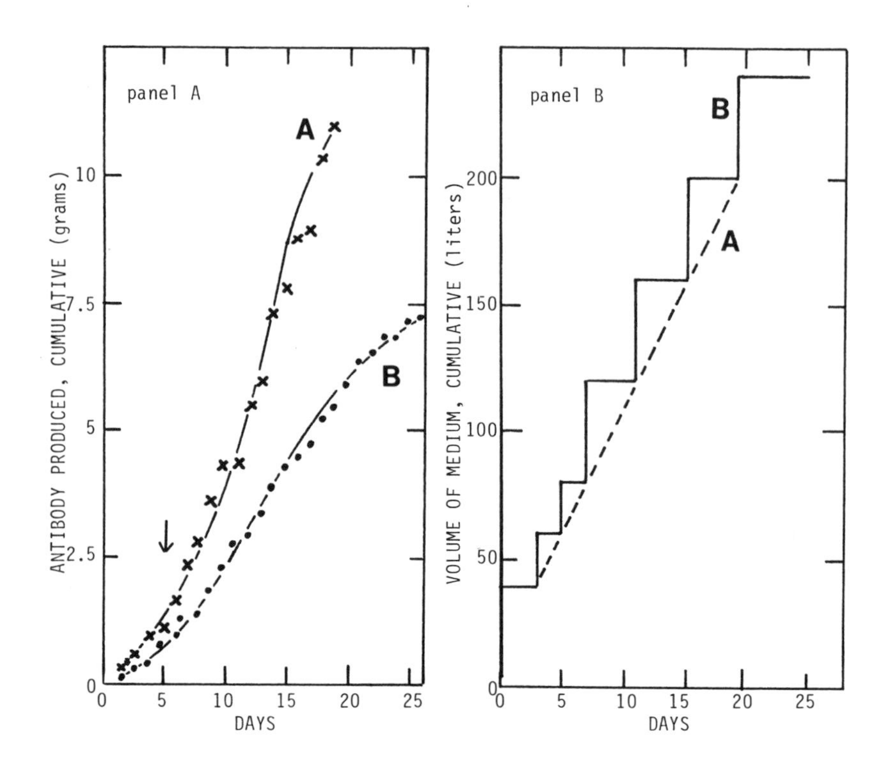

Figure 12. Cumulative antibody produced and medium used by immobilized hybridoma cultures. Panel A: The cumulative amount of antibody produced by the continuously fed culture (x) and the batch fed culture (●) described in Figure 11 is shown. The arrow indicates the time at which continuous feed and harvest was initiated. Panel B: The cumulative volume of medium supplied to the continuously fed culture (dashed line) and the batch fed culture (solid line) is shown.

equilibration of the medium to desired pH, O_2 level, and perhaps nutrient concentration can be accomplished externally to the cells. Also, exchange of medium can be readily accomplished without the need for filtering mechanisms to separate cells from the spent medium. As shown here, this permits the advantages of continuous feed and harvest to be realized. Perfusion of the medium through the immobilized culture also allows for direct and continuous measurements of changes in the medium caused by the cells. An example of the utility of this approach is the use of the oxygen consumption rate in monitoring and regulation of media management.

Other materials such as hollow fibers (7,8) and beds of glass beads (10) also provide for immobilization of cells, and these materials can potentially provide the same advantages as the ceramic. However, problems are encountered with these approaches in obtaining even distribution of cells throughout the support matrix, limiting the potential for scale-up. This may be due to difficulties in effecting homogenous distribution of the cells throughout the support during culture initiation, and to inadequate maintenance of media distribution. Due to the highly uniform configuration of the matrix, these problems have not been encountered with the ceramic.

Due to the potential advantages of the ceramic matrix and the positive results which have been obtained, the system described here has been developed for industrial purposes. Further efforts are focused on increases in scale of the system, and on advances in control and monitoring approaches which have already been shown to be effective in improving the productivity of cell cultures.

REFERENCES

1. Lynn, J.D. and Acton, R.T. 1974. Design of a large-scale mammalian cell, suspension culture facility. Biotechnol. Bioeng. 17: 659-673.
2. Zwerner, R.K., Cox, R.M., Lynn, J.D. and Acton, R.T. 1981. Five year perspective of the large-scale growth of mammalian cells in suspension culture. Biotechnol. Bioeng. 23: 2717-2735.
3. Van Wezel, A.L. and van der Velden-de Groot, C.A.M. 1978. Large scale cultivation of animal cells in microcarrier culture. Process Biochemistry, March: 6-8.
4. Spier, R.E. 1980. Recent developments in the large scale cultivation of animal cells in monolayers. Advances in Biochemical Engineering, 14: 119-162.

5. Tolbert, W.R. and Feder, J. 1983. Large scale cell culture technology. Annual Reports on Fermentation Processes, 6: 35-74.
6. Jarvis, A.P. and Grdina, T.A. 1983. Production of biologicals from microencapsulated living cells. Biotechniques 1: 22-27.
7. Knazek, R.A., Guillino, P.M., Kohler, P.O. and Dedrick, R.L. 1972. Cell culture on artificial capillaries: an approach to tissue growth in vitro. Science, 178: 65-66.
8. Ku, K., Kuo, M.J., Delente, J., Wilde, B.S., and Feder, J. 1981. Development of a hollow-fibre system for large-scale culture of mammalian cells. Biotechnol. Bioeng. 23: 79-95.
9. Lydersen, G., Pugh, G., Paris, M., Sharma, B. and Noll, L. 1984. Ceramic matrix for large scale animal cell culture. Biotechnology. In press.
10. Griffiths, J.B., Thornton, B, and McEntee, I. 1982. The development and use of microcarrier and glass sphere culture techniques for the production of Herpes Simplex virus. Develop. Biol. Standard. 50: 103-110.

APPLICATIONS OF THE MASS CULTURING TECHNIQUE (MCT*) IN THE LARGE SCALE GROWTH OF MAMMALIAN CELLS

Peter C. Brown
Maureen A. C. Costello
Robert Oakley
James L. Lewis

Bio-Response, Inc.
Hayward, CA.

INTRODUCTION

Interest in large scale culture of mammalian cells and the recovery of secreted products from these cells has increased dramatically within the last few years. While this renewed interest initially focused on monoclonal antibody production from hybridomas, an increasingly greater interest in large scale mammalian cell culture has been shown by biotechnology companies whose expertise lies in the area of recombinant DNA. In this latter area, it has become apparent that the favored bacterial and yeast expression systems do not always effect the post translational modifications that are required by certain mammalian proteins for full biological activity whereas expression of these same genes in mammalian cell systems results more often in fully functional protein.

The Mass Culturing Technique (MCT*) was developed to grow the wide variety of mammalian cells required by the research pharmaceutical and biotechnology communities. While the exact configuration of the cell growth chambers varies according to the requirements of each cell type, certain key features are common. These key features include: freshly isolated bovine lymph supplementation instead of serum; continuous growth and

*Trademark of Bio-Response, Inc.

removal of secreted products; mass transfer of nutritional components and metabolic waste products through defined, semipermeable membranes; and modular assembly of cell growth chambers, each of moderate size. The advantages of the MCT system include an unlimited supply of protein supplement for the growth medium; high cell density coupled with low bovine protein content resulting in high initial product purity; rapid re-equilibration with altered media components; and the economic benefits of extended continuous growth.

A. Description of the MCT System

A schematic diagram of the MCT* system is shown in Figure 1. The major components of the system include the bovine lymph based medium, the cell growth chamber with both oxygenation and control circuits and product concentration and purification.

1. Bovine Lymph Based Medium

The thoracic duct of healthy Holstein steers or heifers is cannulated under sterile conditions while the animal is under a general anesthetic. Through experience with over 75 animals, most successful cannulations result from calves in the 400-700 lb range. While lymph flows immediately after surgery, it is not collected for 72 hours in order to eliminate systemic antibiotics which are given just prior to surgery.

The lymph flows through tubing to a radiation sterilized lymph "board" which pumps the lymph through a filtration system at a rate which does not exceed the rate at which lymph leaves the animal. The filtration of the lymph fluid occurs through hollow fiber assemblies with various filtration properties. The lymph filtrate is stored in sterile containers whereas the lymph retentate containing cellular particulates and higher molecular weight components of lymph is discarded. The duration of successful cannulation varies between animals, but some animals have been kept on line in excess of 70 days during which time they have produced over 1500 liters of unfractionated lymph. Throughout, animals are monitored to ensure that blood chemistries and vital signs do not fall below values which are consistent with good overall health.

The exact protein composition of lymph varies depending on molecular weight cutoffs of the filters used in preparing the lymph filtrates. For example, filtrates prepared from the lowest molecular weight cut off filters that we use are essentially devoid of bovine immunoglobulins while retaining representative lower aggregate molecular weight species. As would be expected, filtrates prepared from higher molecular weight exclusion filters include a greater proportion of higher aggregate molecular weight species in addition to the lower molecular weight species. From experience with over 25 cell lines, all but 3 cell lines could be successfully adapted to grow in medium supplemented with at least one of our lymph filtrates. For those lines which would not adapt totally to lymph, supplementation of lymph based medium with 1% fetal serum was sufficient to promote continuous growth.

The rationale for using ultrafiltered lymph stems from an interest in growing cells in a complete medium containing the lowest possible content of bovine proteins. In this way, the task of concentrating and purifying the secreted products becomes much easier to accomplish. Regarding actual cellular production rates in lymph based medium versus medium supplemented with fetal calf serum, no significant differences in monoclonal antibody production were seen when one cell line which grew equally well in either medium was tested for specific immunoglobulin production in static cell culture. An additional benefit of the ultrafiltration process is that bacteria or viruses, if any, carried by the whole lymph would be physically removed from lymph before the lymph enters the cell growth chambers. We are currently validating representative fiber cartridges in order to support this hypothesis. Clearly, protein supplement free of bacteria or bovine viruses is highly desirable if the cellular products produced in this material are to be used therapeutically.

While filtered lymph may be transported to the cell growth chambers directly, it is more appropriate from the standpoint of process control to interrupt the flow of lymph and accumulate daily pools of lymph fluid which are quarantined. These pooled batches are tested for total protein content, relative protein composition, for the absence of bacterial contaminants, and are subsequently released for use if the test results meet established criteria.

For the preparation of the complete cell growth medium, the appropriate lymph filtrate is blended with conventional cell culture medium at predetermined ratios and introduced into the cell growth chambers. While more specialized media may be used, a 1:1 mixture of Dulbecco's Modified Eagles Medium and F-12 has been found satisfactory for all but 1 cell line tested.

2. Cell Growth Chambers

The main characteristic of the cell growth chambers is that the transfer of nutrients occurs through semipermeable membranes in a hollow fiber configuration. In Figure 1, the schematic shows that the cells are essentially bounded by 2 membranes: the Cell Exclusion Membrane (CXM) and the Product Exclusion Membrane (PXM). The CXM is a microporous membrane with pores in the 0.1-0.2 μM range whereas the PXM is an ultrafiltration membrane with a nominal molecular weight cutoff chosen to retain the product and the vast majority of the proteins contained in lymph.

Oxygenation occurs through a circulating loop directly adjacent to the CXM and relies on the diffusion of oxygen rich medium created in the oxygenator into the cell growth chamber. Peristaltic pumps are used throughout. The resulting pulsed delivery of fluid from these pumps seems to facilitate the passage of oxygen rich medium across the CXM. The oxygenators themselves are silicone rubber based and are either available commercially or are of our own design. For all but the highest cell densities achieved in the system (approximately 2-4 x 10^7 cells/ml) compressed air with added CO_2 contains sufficient oxygen to maintain the partial pressure of oxygen in the growth chamber above 30-40 mM of Hg. For additional oxygenation, both the pumping rate through the loop and oxygen partial pressure in the gas used to equilibrate the medium in the oxygenator may be increased.

Opposite the oxygenation loop is shown an additional fluid circuit through which the majority of pH control occurs. Diffusion across the PXM removes excess lactic acid and other low molecular metabolites and replenishes vitamins, cofactors, and amino acids which have been depleted within the cell growth chamber. Lacking this circuit, increased throughput of protein supplemented medium would be necessary to maintain the cells in a chemostat mode. The net flow through this control circuit is regulated in order to keep lactic acid levels below approximately 15 millimolar and the medium above pH 6.8 .

The cell growth chamber, which is in effect the extra capillary volume of a hollow fiber cartridges, does not exceed 1 liter in our largest growth chamber which is capable of growing anchorage independent cells. Protein supplemented medium is perfused directly into the chamber and is withdrawn along with cell free secreted products through the CXM. Further access to the growth chamber through the excess cell drain allows direct sampling of the cells and media within the chamber itself. For suspension cultures an important aspect of the growth chambers is the ability to remove cells from the chamber. We have found, for example, that cell populations which are maintained in the range of 2-4 x 10^7 cells/ml by periodic removal of cells results in a population with greater viability as determined by dye exclusion than populations that are allowed to exceed that number. Culture units with greater than 5 x 10^7 cells/ml become difficult to manage as oxygen tension drop to near zero and the pH falls below 6.8. In addition to the metabolic demands of cells at high density, significant cell packing may occur which further aggravates mass transfer through the semipermeable membranes. The tendency for the cell packing varies widely, however, among cell types. As a result, some cell lines are more amenable to high density culture than are others (e.g. cell densities in excess of 2 x 10^7 cells/ml). Regardless of the cell type, the entire growth chamber is rotated through 360 degrees in an acentric manner with a period of about 20 seconds in order to discourage settling of cells and thus minimize the packing problem.

For anchorage dependent cells, i.e. cells that do not adapt to growth in suspension culture, the cell growth chamber is modified so as to include a large surface area which is composed of a material suitable for the attachment of anchorage dependent cells. While a number of materials have been tested, the majority of our experience in this area is with glass beads 1-3 millimeters in diameter (1). In contrast to suspended, neutral density microcarriers in stirred tanks (2), glass beads are allowed to settle and result in a packed bed configuration through which oxygenated medium is percolated. While the surface area of the packed bed is less than that achievable with microcarriers, the advantage of the packed glass bed are low cost, reusability, and ease of scale up.

Finally, all culture unit are extensively monitored through the analysis of samples removed from the sampling ports indicated by the letter A in Figure 1. Critical measurements of oxygen tension, lactic acid concentration, cell count and viability (where applicable), pH and product assays provide the data upon which decisions relating to culture unit management are made. In addition, lactic dehydrogenase (LDH) provides an indirect measure of the relative health of the cell population within each culture unit inasmuch as LDH activity is an intracellular enzyme, the activity of which increases in the culture fluid as cells become necrotic.

3. Product Concentration

Cell free supernatant from the growth chamber is collected continuously in a cold room adjacent to the 37 degree Celsius culture room. At various intervals the pooled product is

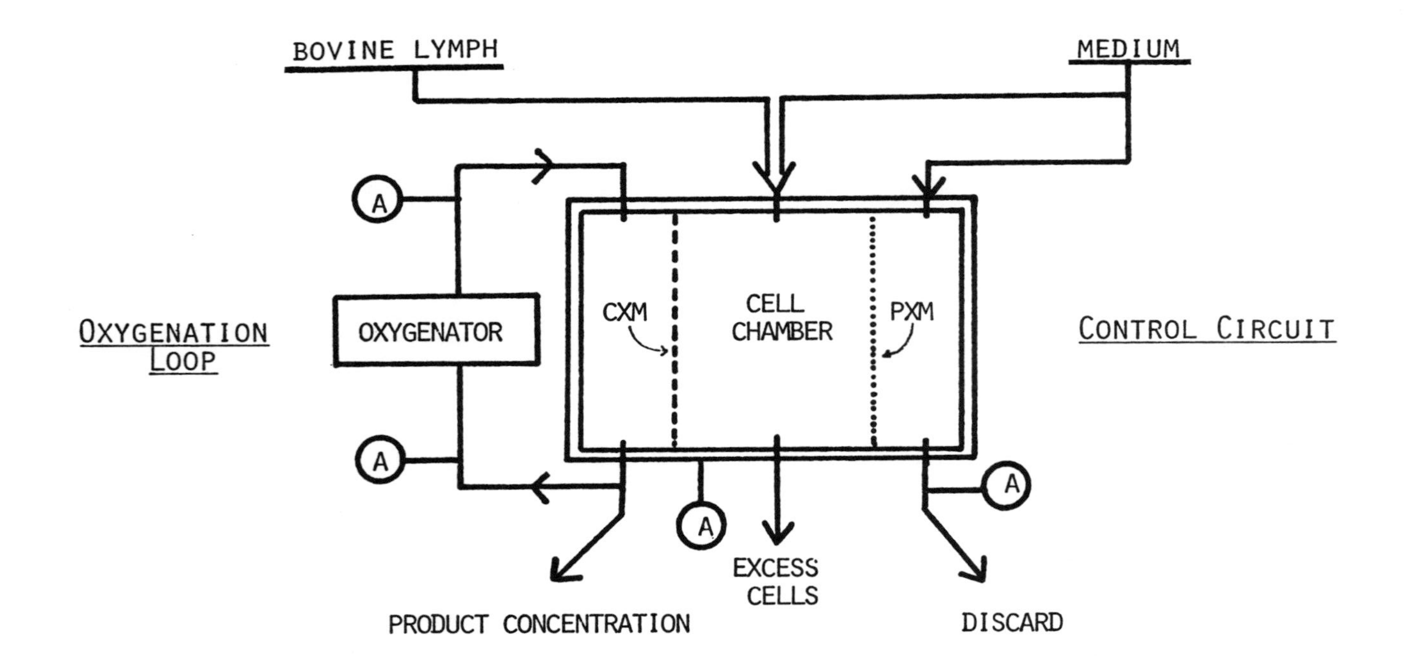

Fig. 1. Schematic diagram of the MCT* system. Symbols used: CXM, cell exclusion membrane; A, sample ports.

removed and sterility is maintained while product is concentrated over hollow fiber cartridges a minimum of 20 fold, depending on specific product requirements.

B. **Results**

1. Cells Grown in the MCT System

A summary of cell grown to production levels in the MCT system is shown in Table I. As may be seen, a wide variety of cell types have been successfully accommodated. Regarding the production of monoclonal antibodies, over 130 grams have been produced from a total of 10 cell lines representing the three most commonly used mouse fusion partners. Moreover, the antibody produced consisted of three of the most common isotypes, IgG1, IgG2A, and IgG3. In addition, two of these cell lines were grown in medium supplemented, not with lymph, but with 5% fetal celf serum in one case and in commercially available, serum free medium in the other. Thus, while the system itself was designed to utilize bovine lymph, it is not dependent on lymph and may function well with alternate sources of protein supplementation.

Of the cells other than hybridomas grown in the system, the majority are anchorage dependent and many had been genetically modified to secrete human protein. Two Chinese hamster cell lines were grown both as suspension and as anchorage dependent cells. Because of the nature of the specific plasmid constructs used to introduce the human genes coding for these

Table I. Cells grown in the MCT system

Cell Type	Species	Comments
Hybridoma (5)	mouse	Sp/2 fusion partner
Hybridoma (3)	mouse	NS-1 fusion partner
Hybridoma (1)	mouse	P3x63Ag-8 fusion partner
Hybridoma (1)	rat	Y3 fusion partner
Lymphoblastoid	human	
Kidney *	human	primary cell line
Melanoma *	human	genetically engineered
Chinese hamster (3)*	hamster	genetically engineered
Fiborblast *	mouse	genetically engineered

- Number in parentheses indicates number of cell lines grown of that type.
- Asterisk (*) indicates anchorage dependent cell lines.

proteins, specific inducing or selecting agents were continuously infused into the system along with medium and lymph to maintain high levels of expression. In addition, the continuous flow of nutrient to 2 other cell lines grown in the MCT system allowed us to gradually reduce the protein supplement as cells moved from a growth or expansion phase into a less mitotically active production phase. The longest production run for 1 anchorage dependent cell line was over 120 days; the average run was about 70 days.

2. Hybridoma Growth and Monoclonal Antibody Production

The results of a monoclonal antibody production run lasting 50 days is shown in Figure 2. The cell growth chamber itself was a hollow fiber cartridge assembly which included an extracapillary volume of about 300 ml. Cells (2×10^8) were innoculated into this extracapillary space at day zero. After After a lag period of about 10 days, cell numbers increased rapidly to about 1×10^7 viable cells per ml (panel B). Cell counts, however, are subject to significant underestimation because of our inability to effectively sample the entire growth chamber and the inherent tendency for cell to aggregate. Thus, estimates of cellular productivity are suspect based on these data. Whereas measured cell counts began to decline after day 30, productivity continued to increase (panel A) until plateau values of about 600 mg/day were achieved after day 40.

As may be seen in Panel D, lactic acid levels were maintained well below 10 millimolar. We have found that lactic acid levels are good monitors of cell metabolism (3) and, in addition, measure the extent to which the growth medium has been depleted. In order to maintain only moderate levels of lactic acid, a total of about 250 liters of medium and 110 liters of lymph were utilized to produce 15 grams of antibody over the course of the production period. Because of the arrangement of our growth chambers, actual product was recovered in a volume that was less than these figures would indicate. Improved media utilization has been achieved recently as the levels of lactic acid have been allowed to approach 15 millimolar. For comparison, an exhausted medium from static culture with a hybridoma cell population of about 1×10^6 cells/ml may have lactic acid levels in excess of 25 millimolar.

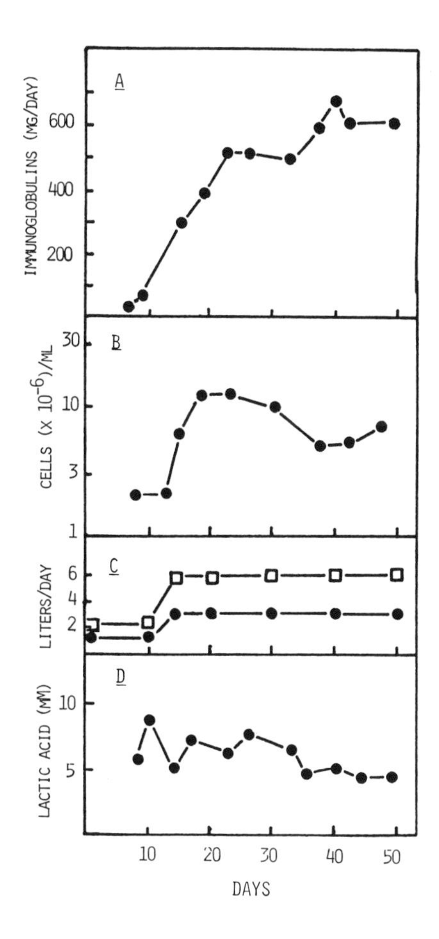

Fig. 2. Production of a mouse monoclonal antibody in the MCT* system. A mouse hybridoma cell line (P3x63Ag-8 fusion partner) which secreted an IgG2A monoclonal antibody was innoculated into an MCT growth chamber (approximately 300 ml culture volume) and grown for 55 days in antibiotic-free medium. Panel A, daily production rate of mouse immunoglobin determined by ELISA assay on cell free supernatants removed from the cell growth chamber. Panel C, daily utilization of lymph and tissue culture medium: -•-, lymph filtrate; -□- medium. Panel D, lactic acid concentrations from samples withdrawn from the cell growth chamber.

3. Production from a Genetically Engineered Mouse Fibroblast Cell Line

The packed glass bead growth chamber was chosen for the growth of an anchorage dependent mouse fibroblast line because of the superior performance of this culturing unit when compared to the performance of a polysulfone based hollow fiber unit (4) with a comparable surface area of 1 m^2 (data not shown). As may be seen in Figure 3, production of the protein which resulted from the introduction of the gene coding for that hormone continued to increase until production of about 4 mg per day was achieved. For comparative purposes, this level of production was approximately equivalent to the daily production of this same cell line at confluency in about 50 roller bottles (890 cm^2) or about 2 x 10^{10} cells.

Over the course of the production period approximately 100 mg was collected with the total expenditure of about 250 liters of culture medium and 60 liters of lymph filtrate. As in Figure 2, the lactic acid levels were maintained at about 5 mM, a figure which from our experience results in a vigorous population of most cell types.

4. Purity of Monoclonal Antibody Produced in the MCT System

Because selected lymph filtrates are used in the growth of most hybridomas, it is possible to achieve levels of purity which exceed 50% with respect to total protein. Furthermore, it is also possible to achieve concentrations of antibody which exceed 150 mg/L as a result of the cell density maintained in the MCT system. The advantage in purification of antibodies of such high initial purity may be seem in Figure 4 which is a reducing SDS polyacrylamide gel of a mouse IgG2A produced in a low molecular weight filtrate of lymph which had been supplemented with human transferrin at 10 μg/ml. In this latter regard, we have seen that the addition of transferrin to low molecular weight lymph filtrates is helpful in developing full growth promoting activity. Lane 2 shows concentrated cell free supernatant removed directly from the cell growth chamber. This preparation was judged to be over 50% IgG2A by comparison with total protein in the preparation. The predominance of the heavy and light chains is apparent. Of additional interest is that after ammonium sulfate precipitation (lane 3) and cation exchange chromatography (lanes 5,7) the antibody preparation is over 95% pure with respect to contaminating bovine protein.

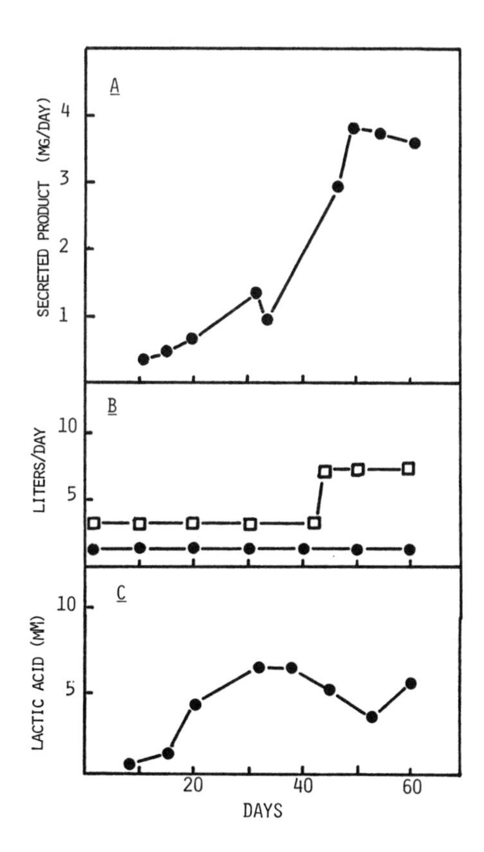

Fig. 3. Production of a genetically engineered protein from an anchorage dependent mouse fibroblast cell line. Cells (4×10^8) were introduced into a packed glass bead culture unit which had approximately 1 m^2 of surface available for cell attachment. Panel A, daily production by ELISA assay of the human protein secreted by these cells. Panel B, daily utilization of lymph and tissue culture medium: •, lymph filtrate; □ medium. Panel C, lactic acid concentration in samples removed from the cell growth chamber.

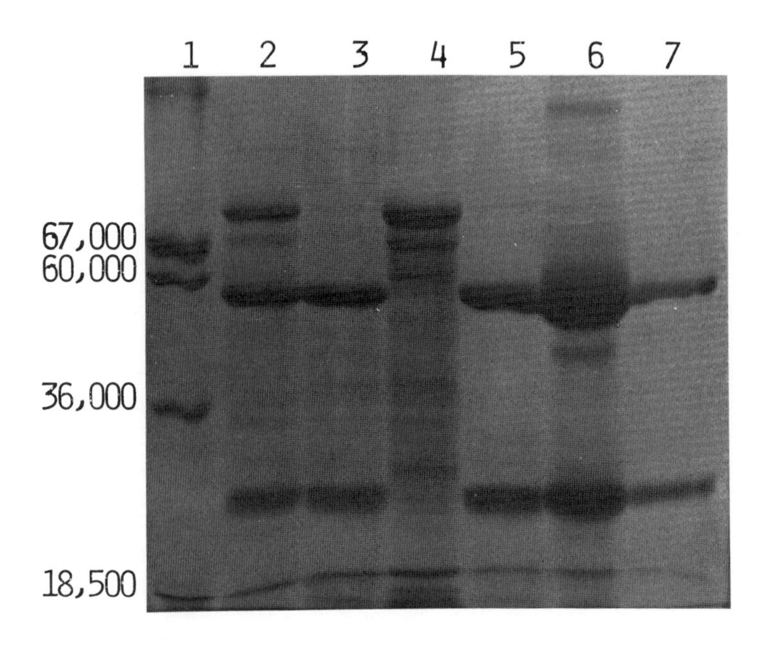

Figure 4. SDS-Polyacrylamide gel analysis of a mouse monoclonal antibody produced in the MCT System. Analysis was performed using standard Laemmli SDS-PAGE, 3% stocking gel, 10% running gel. Lane 1, Pharmacia high molecular weight standards (reduced). Lane 2, 30-fold concentrate of eluate from a culture unit. Major bands are transferrin, heavy chain and light chain of the antibody. Lane 3, an aliquot of material in lane 2 was precipitated using 0-30% and 30-50% final concentrations of ammonium sulfate. Material in lane 3 represents redissolved and dialyzed protein from the 30-50% precipitate. Lane 4, proteins remaining in the supernatant after a 30-50% ammonium sulfate precipitation. Lane 5, and 6, the material in the 30-50% ammonium sulfate precipitate was dialyzed extensively against low salt buffer pH 7.8-8.8 and applied to two different DEAE ion exchange columns. Lane 5, protein in pass-through of column (pH 8.8 at 4 degrees C). Lane 6, a gradient elution of an aliquot of the same preparation, (pH 7.6 at 4 degrees C).

Lane 7, final pool of antibody after a 30-50% ammonium sulfate precipitation and a single pass of precipitated material over a DEAE ion exchange column.

SUMMARY

We have described a system which is capable of growing and recovering the secreted products of cell lines cultured either in suspension or attached to defined substrates. Freshly prepared bovine lymph substitutes for serum and offers an advantage over serum because of the abundance of lymph that can be obtained and the resulting cost advantage of this lymph versus serum. In addition, ultrafiltration of lymph results in medium which is diminished in contaminating bovine protein and is largely, if not totally, pathogen free. Suspension cells are grown continuously with the perfusion of lymph supplemented medium through semipermeable hollow fiber which radiate throughout the cell growth chamber. Anchorage dependent cells are grown in packed beds of glass beads. Both systems offer the cost benefits of continuous of growth.

ACKNOWLEDGEMENT

The authors acknowledge the pioneering work of Sam Rose in the development of the MCT* system.

REFERENCES

1. Varani, J., Dame, M., Beals, T., and Wass, J. (1983), Biotechnology and Bioengineering 25: 1359-1372.
2. Levine, D., Wong, J., Wang, D., and Thilly, W. (1977) Somatic Cell Genetics 3: 149-155.
3. Cristofalo, V. and Kritchevesky, D. (1965) Proceedings of the Society for Experimental Biology and Medicine 118: 1109-1113.
4. Knazek, R., Guillino, P., and Kohler, P. (1972) Science 178: 65-67.

PANEL DISCUSSION 1

DR. PARCELLS (Upjohn, Kalamazoo, MI): I am directing my question to the last speaker, Dr. Brown. Do you find any difference in the species of cows as far as quality or composition of the lymph? I noticed in your photograph that it looked like Holstein's. Have you ever tried Guernsey?

DR. P. BROWN (Bio-Response, Inc., Hayward, CA): Earlier, we tried different types of cows. Holsteins have been bred for docility..., they stay in those standrions the entire time that they're on duty. So we have been using Holsteins exclusively. Right now we are working in the 600-700 pound range but large ones don't provide the continued flow. Animals are kept on for as long as 80 days producing 10-20 liters per day of lymph.

DR. C. HO (State Univ. of NY at Buffalo, NY): I want to ask Dr. Lydersen about the oxygen electrode. All of these are notoriously sheer sensitive; i.e. with one flow rate you get one concentration, with another flow rate another concentration. How did you solve this problem?

DR. B. K. LYDERSEN (Research and Development Laboratory, K C Biological, Lenexa, KA): In practice, we use a probe chamber which is a glass vessel with a volume of 50 ml in which the probe sits. We tested that earlier and did not observe any problem. We varied the flow rate from 50 ml/min up to 2 liter/min and saw only an inconsequential change in the reading. The one thing that can happen is that you pressurize the system as you increase your flow rate which has a small effect, but we have not seen that problem and have tested for it.

DR. C. HO (State Univ. of NY at Buffalo, NY): Based on some of your consumption rates by a simple calculation you will find that if you have a residence time of, for example, one minute, then your oxygen consumed through that loop is somewhere around .01 milligrams. Will your electrode be able to detect such a small change?

DR. B. K. LYDERSEN (Research and Development Laboratory, K C Biological, Lenexa, KA): The whole logic is that this is a continuous process. The probes are there all the time, and the oxygen consumption rate in the culture doesn't change much in time. Over an hour, the most you'll see is a few units difference. Consequently, you don't see rapid changes and it's not necessary to detect them. I think you're right. If there are rapid changes going on in terms of major differences over a few minutes, we won't detect that accurately.

DR. C. HO (State Univ. of NY at Buffalo, NY): I would like to address a question to Dr. Rupp. Regarding encapsulated cells, you say the size is somewhere around 500-800 microns in diameter. Do you ever observe any oxygen starvation in the core of the capsules? In other words, is there any cell degeneration occurring?

DR. R. G. RUPP (Damon Biotech, Inc., Needham Heights, MA): We cannot and have not, as yet, measured the actual oxygen concentration inside the microcapsule. I was asked after my discussion what was actually happening, what terminated these cultures. In point of fact, the viability of the cells after day 10 or 12 is beginning to drop and we feel it is because there is a limitation, maybe not in just oxygen, but in other nutrients getting to the core of the cells inside the microcapsule. There is a large cell mass there. So I think something is rate-limiting and it may be oxygen as well as other factors.

DR. C. HO (State Univ. of NY at Buffalo, NY): If the specific cell uptake rate is similar to that of bacteria, then it only would penetrate about 100 microns. That could be a problem.

DR. R. G. RUPP (Damon Biotech, Inc., Needham Heights, MA): As I said, our goal is to make a 100-200 micron capsule to attempt to solve that. We're just happy that we're getting the production we're seeing with these large capsules. When we go to the smaller capsules, we have seen significant increases in the total viable cells in the microcapsules.

DR. B. C. BUCKLAND (Merck and Co., Robway, NJ)): This question is directed to Dr. Birch. I was very interested in the slide for the continuous culture data where you indicated that the production levels under oxygen limitation conditions seem to be similar to the other conditions. Is that a typical result or have you only looked at one particular process?

DR. J. R. BIRCH (Celltech Ltd., Slough, Berkshire, United Kingdom): That seems to be pretty typical, yes.

DR. B. C. BUCKLAND (Merck and Co., Robway, NJ): There is a slide illustrating some chemostat data, and the productivity under oxygen limited conditions seems to be similar to other conditions illustrated there. I was very curious about that and wondered if it was a typical result. Do any of the other speakers have comments on that?

DR. J. FEDER (Monsanto Company, St. Louis, MO): I believe that slide was dealing with glutamine, glucose and oxygen.

DR. J. R. BIRCH (Celltech Ltd., Slough, Berkshire, United Kingdom): Yes, that is correct. Overall they were broadly similar.

DR. B. K. LYDERSEN (Research and Development Laboratory, K C Biological, Lenexa, KA): What intrigued me was the approach of limiting the culture as opposed to the opposite approach, which would be to allow the maximum cell growth which seemingly would allow you to increase the feed rate or the production rate. It doesn't seem obvious why you would want to limit the culture the way you did.

DR. J. R. BIRCH (Celltech Ltd., Slough, Berkshire, United Kingdom): You have to control the culture with something. You can either do it by limiting a nutrient; you could do it as some people have done. You can run it as a cytostat by regulating it on a constant cell number. You could regulate it on the concentration of inhibitory metabolite, for instance. There are various ways which you could regulate it but you have to have something which can regulate the steady state. Perhaps the slide I showed was slightly misleading in that the cell population we worked with there, about a million cells per milliliter was quite low. We were deliberately working with low populations to be absolutely sure we were in the nutrient limitation range we wanted to study. We can get very much higher cell populations of by appropriate design of the media, in fact we can and have worked at cell populations at least ten times as high as that.

DR. J. HOPKINSON (Amicon Corporation, Danvers, MA): I have a comment and then a question for Dr. Lyderson. With regards to these distances for mass transfer, we have found in very early experiments with hollow fiber culture systems that the packing density as we call it, but basically the inner fiber spaces and the spaces between fibers are very critical with regard to keeping cells alive. In our systems we have packing densities which result in 100-200 microns as the maximum distance between the perfusion flow and the most distant cells so that 100-200 micrometer distance is fairly critical. The question I had for Dr. Lyderson with regard to the "B" type culture cartridge that you have, do you have any data that gives cell density either in centimeters squared or per mls of medium for immobilized cells in those culture systems relative to suspension systems. In other words, you presented a great deal of data showing cell densities in the "A" types cultures somewhere in the range of 1 or $2X10^5$ cells/cm^2. What number corresponds to that in the immobilized cells in the "B" cartridge?

DR. B. K. LYDERSEN (Research and Development Laboratory, K C Biological, Lenexa, KA): I guess for the "B" type we made a rough calculation of what we felt was the number of the hybridoma cells. Those cartridges contain roughly a liter of medium. So that would calculate to about $5X10^7$ cells/ml. Now, of course, the medium is being constantly perfused so the system turns out to have about $5X10^6$ or less cells per ml of total medium circulated.

DR. J. FEDER (Monsanto Company, St. Louis, MO): What is the residence time in the reactor?

DR. B. K. LYDERSEN (Research and Development Laboratory, K C Biological, Lenexa, KA): Basically, for those last experiments which I described, they were all run at 1 liter/min.

DR. J. FEDER (Monsanto Company, St. Louis, MO): That is certainly a very significant perfusion rate, particularly with respect to the question of oxygen transfer. That is, if you recirculate 1 liter per hour of re-aerated medium, the 5×10^{10} cells in the reactor are actually exposed to 60 liters of freshly aerated medium per hour. Under these conditions the change in oxygen concentration would be negligible, minimizing aeration problems.

DR. B. K. LYDERSEN (Research and Development Laboratory K C Biological, Lenexa, KA): Yes, in fact that's what I alluded to in the two approaches of a fixed culture or an immobilized culture versus homogeneous. It allows you to manipulate the medium easily.

DR. B. MAIORELLA (Cetus Corporation, Emoryville, CA): I am directing my question to Dr. Birch. Much of the early literature on cell culture describes the problems of damage to the cells under aeration conditions and aeration rates were limited to less than 0.02 VVM's or else methyl cellulose was added to stabilize the cells. Can you give us an idea of the air flow rates in your system and whether you had to take any steps to protect the cells against bubble damage?

DR. J. R. BIRCH (Celltech, Ltd., Slough, Berkshire, United Kingdom): I'll answer the second part first. No, we don't have to take any specific precautions in terms of VVM. I guess we're typically working at something around 1 VVM, that sort of flow rate.

DR. W. R.TOLBERT (Monsanto Company, St. Louis, MO): I would like to direct my question to Dr. Lydersen. In your ceramic "B" which allows the cells to attach that don't ordinarily attach, have you increased the surface charge density to enhance that and if so, is there any problem with medium components being leeched out of the medium as was the case with some of the earlier microcarriers?

DR. B. K. LYDERSEN (Research and Development Laboratory, K C Biological, Lenexa, KA): The second part of the question is an easy one to answer. We haven't observed any problems with leeching out of medium components. In other words, at least a large amount of binding of transferin or whatever, we don't observe. With regard to the first question, to find the ceramic we really used an empirical approach where we tried different things and saw what happened. We have not defined, specifically, the basis for our success, whether it's charge, entrapment or what. So the only answer I can give you is that we found something that works and does what we want it to. I think over the years we'll define better why this works, but right now it's mainly a matter that it does work.

DR. G. BELFORT (Resnnelaer Polytechnic Institute, Troy, NY): My question is directed to Drs. Lydersen and Birch. Can you tell me a little bit about the shear sensitivity of these hybridomas? They are being pumped around and subjected to different shear rates. I'd be very interested to know how stable they are. For example, if you have them near the impeller and there is high shear gradient in the mixing unit or in your case the air lift system, do you see any problem if you change the velocity of the air lift system in terms of cell damage? This is a general question about cell damage as a function of shear.

DR. J. R. BIRCH (Celltech Ltd., Slough, Berkshire, United Kingdom): I would like to comment on the effects of shear. We don't see problems of mechanical damage to the cells in the air lift reactor and that was claimed to be one of its advantages, that it is a low shear type of reactor. Having said that, I think you're very right to point out the problem of mechanical shear. It is something that's been very little studied. People talk rather glibly about the possibility of shear damage at the tips of impellers, etc., but really there is very little knowledge and certainly very little experimental data on the effects of different types of shear and the quantitative effects of shear on animal cells.

DR. B. J. LYDERSEN (Research and Development Laboratory, K C Biological, Lenexa, KA): Along that line, people talk about shear but there is very little understood. One question that I would have would be underwhat what medium conditions might this be studied. If there's little or no protein, the cells may end up being more shear sensitive, and I know one protein-free medium which was designed for hybridomas that does not work in suspension whereas it does work for flasks. So that suggests that as you lower the protein content of medium, these cells become more sensitive.

DR. J. FEDER (Monsanto Company, St. Louis, MO): There is an obvious case where shear effects are important and that is microcarrier attached cells grown in suspension. The shear effects act to detach the cells from the microcarriers. We have observed some microcarrier grown cells that remained firmly attached as long as they were in a serum-containing medium but which detached when transferred to a serum-free medium, unlike the T-flask grown cells which remained attached in either case.

DR. C. BURLEY (Hoffmann-La Roche): Along these same lines, could you describe the sparges in the airlift fermenter?

DR. J. R. BIRCH (Celltech Ltd., Slough, Berkshire, United Kingdom): Yes, it's a sparge ring on the base of the draft tube with just a few relatively large holes, each with a diameter of about 1.5 millimeters.

DR. J. FEDER (Monsanto Company, St. Louis, MO): Dr. Lydersen showed a cell growth curve in which the cells were approaching peak concentrations, but the oxygen consumption rates were leveling off. There was some question as to what the oxygen consumption rate should have been at that cell density. One might suggest that the cells were entering a nonproliferative state and the oxygen requirement of such cells could be significantly different from cells in log growth. Perhaps this occurs once the cells have reached confluency in these reactors.

REGULATORY CONSIDERATIONS FOR PRODUCTS DERIVED FROM THE NEW BIOTECHNOLOGY*

John C. Petricciani

Food and Drug Administration
Bethesda, MD

Recent advances in biotechnology have made it possible to move at a very rapid pace from laboratory research findings to potentially useful drugs and biologicals. Although recombinant DNA (R-DNA) work started with bacteria as the producer cells, it became clear relatively early that bacteria would not be able to synthesize certain molecules. For example, hepatitis B surface antigen is not produced by recombinant bacteria in sufficient amounts to make it economically feasible. Another deficiency of bacteria is their inability to glycosylate. If biological activity of a macromolecule depends on the presence of one or more glycosylations, then bacteria could not be used. As a result of these shortcomings of bacteria, there has been increasing interest in the use of mammalian cells for R-DNA work. The other major area of research and development involving the new biotechnology is, of course, monoclonal antibodies produced from hybridomas. Large scale cultivation of mammalian cells has also made it possible to prepare improved versions of older products such as inactivated poliovirus vaccine. In each of these three

---

*Based on a presentation at the Workshop on Abnormal Cells, New Products, and Risk; held at the National Institutes of Health on July 30-31, 1984.

areas the mammalian cell substrate has been the central focus of interest with respect to safety issues because the cells being used or proposed for use are abnormal, and in fact some of them are frankly tumorigenic.

The history of what is and what is not acceptable with regard to cell substrates when they are being used to manufacture a product for human consumption goes back to the early days of vaccine development. In 1954 a decision was made that human tumor cells were unacceptable, and that primary cultures derived from normal tissues of normal animals should be used. That decision set the tone for discussions on the subject up through the present time. The criteria of acceptability for cell substrates were discussed numerous times during the decades following that initial landmark decision, including several major international meetings. Eventually the concept of human diploid cell lines was accepted, but only after long and vigorous debates. The last five years, however, have provided a literal explosion of products derived from other than primary or diploid cells. As a result, the international biomedical community has had to begin to take some serious second looks at the bases on which cells might be considered acceptable in the manufacture of various products.

The generic question regarding the use of abnormal cells is: is it safe to use products derived from them? In order to begin to answer that question, one has to ask how risk would be transmitted from the cells to the product. The table below lists the possible routes by which one or more cellular factors could pose safety issues in a product. As a first step, one would like to have assurance that the product contains no viable abnormal cells. Although it is conceivable that in the future a product could undergo so little processing and purification that live cells could be a realistic concern, those are not practical concerns for current products.

FACTORS ASSOCIATED WITH POTENTIAL RISK FROM CELLS

Factor	Effect
• Cell	• Tumor
• Proteins	• Immune Response • Transformation
• Nucleic Acids	• Integration and Expression of Abnormal Gene
• Endogenous Viruses	• Transformation

Fig. 1

Subcellular factors are of much more than theoretical significance, and must be considered on a case-by-case basis for each cell system and each product. In some cases certain factors will be irrelevant, or of minor significance, in comparison to others. Contaminating cellular proteins, for example, could, if present in high enough concentration induce antibodies. And if those happened to involve histocompatibility antigens, it would be desirable to reduce their level. Fortunately, this has not been a practical problem so far.

Residual cellular DNA (RC-DNA) is quite another matter, however. As more and more serious consideration has been given to the use of abnormal mammalian cells in the manufacture of products, the safety issue has focused more and more on the amount of RC-DNA in the products. Basic to that focus of concern is the proposition that whatever abnormal biological characteristic one is concerned about transferring in the product it has its roots in the DNA of the cell. Specifically, the safety issue is concerned with the possibility that RC-DNA, if present in sufficient quantity and with biological integrity, could transmit the genetic information for those abnormal characteristics to the human recipients of the products. The problem is complex and the data available to help answer the questions have not been extensive. It is complex because we do not know how most of those abnormal characteristics fit together in a genetic sense or for that matter how some of them interrelate at the level of phenotypic expression. Is infinite life potential, for example, independent of tumorigenicity, invasiveness, metastatic potential, or soft agar colony forming ability? How do each of those characteristics, and others, interrelate? Much of that is as yet unanswered. As a result, the approach to the problem has had to be somewhat indirect, but based on pragmatic considerations.

Since there is a long history for the use of normal mammalian cells in the production of vaccines, and they have, to some degree or another, all contained a certain amount of RC-DNA about which there has been no concern expressed or known adverse effects, we already have a baseline of sorts above which one would not like to go for products derived from abnormal cells. Because only a small fraction of the total genome of abnormal cells is different from that of normal cells, it is only the abnormal DNA about which there should be concern. In other words, since the presence of RC-DNA from normal cells has already been accepted, it is only the abnormal DNA of abnormal cells about which there should be special concern. Although oncogenes have received a great deal of attention during the past several years, they almost certainly will not be sufficient to explain the entire process of tumorigenesis. There are other genetic factors such as strong promoters, gene amplification, gene activation, and gene suppression which at least in some cases will be of significance. However, cellular oncogenies do represent specific sequences of some tumor cell DNAs which have transforming potential; and in that sense they are identifiable pieces of abnormal DNA from abnormal cells. As such, they can be useful in trying to assess the biological significance of various levels of RC-DNA.

As an example of how one might approach this assessment, one can bring together several pieces of data to calculate how far removed one might be from a measurable biological effect by RC-DNA. Because of efforts by industry, many of the new biotechnology products contain about 10 pg of RC-DNA per dose or less. This is equivalent to about one mammalian cell genome. Because the amount of RC-DNA in the final product is so low, it is not technically feasible to determine its size distribution. A reasonable assumption would be that RC-DNA contains random portions of the genome with various sizes being represented, and that at least some of the sequences are large enough to retain biological activity. There is no reason to assume either selective elimination or selective concentration of any sequences. As a worst case, one could assume that the cell substrate had an oncogene in each cell, and that a single cellular oncogene was present in an undegraded form in the 10 pg of RC-DNA. The smallest cellular oncogenes are in the range of 1 kb which is equivalent to about 10^{-6} pg of DNA. In order to relate that amount of abnormal DNA to a biological event involving risk, one can look again at a worst case situation. *In vitro* studies using optimal experimental conditions have shown that various purified cellular oncogene DNAs have a transforming efficiency of about 10^4 focus forming units per ug. That is equivalent to 100 pg per transformation focus. The difference, then, between a minimally effective dose of transforming cellular oncogene DNA *in vitro* and the amount that might be in a product is the difference between 100 pg and 10^{-6} pg, which is 8 logs. While there are a number of assumptions in this analysis, it is important to recognize that most of them are heavily weighted in favor of a transformation event, and that if one were to be more realistic in the assumptions by taking into consideration the biological barriers to uptake and expression of exogenous DNA *in vivo*, the 8 log difference would not diminish, but rather become larger.

This analysis of the issue was presented recently at a workshop on the use of abnormal cells in the production of new products. In addition, various other experimental data were presented and discussed in order to try to arrive at a consensus on the issue of risk associated with RC-DNA in products derived from abnormal cells. The overall conclusion from that meeting was that there is sufficient evidence now available to consider an abnormal cell acceptable under two circumstances which are not mutually exclusive. First of all if the manufacturing process consistently results in a product with the amount of DNA per dose in the picogram range, then it should be safe. That was essentially a quantification approach which assumes that there may be sufficiently large pieces of DNA to be biologically active, and relies on reduction to such low levels that they become of no real consequence. The other approach which was considered and preferred by many, was to see if the manufacturing process for a given product would result in the loss of detectable biologically active DNA in the final product. That, of course, is the most direct way of getting at the question of safety because ultimately one wants to know whether or not the final product has biologically active DNA in it. If the manufacturing process can be shown to eliminate that activity, then the quantification of RC-DNA becomes of much less importance. Using this approach leaves the door open with regard to the amount of RC-DNA that might be acceptable in a final product. The situation becomes somewhat more complicated, however, if the product is an intact viral vaccine because of the possibility of incorporating cellular DNA into the virion itself and of then transferring it to humans in a very much more efficient manner than if it were free DNA. As a result, there are two additional questions which need to be answered for viral vaccines produced in abnormal cells. The first, of course, is whether or not cellular DNA is incorporated into the virus. If the answer is no, then the issues are the same as those for non-replicating agents such as interferon. But even if it were shown that cellular DNA was incorporated into the virions, one could continue to consider the vaccine if it were inactivated and the cellular DNA in the virions could be shown to be inactivated along with the viral DNA or RNA. Obviously, those data are going to be considerably more difficult to generate than the quantification of RC-DNA.

The other major safety issue relating to the use of abnormal cells is the presence of endogenous and exogenous viruses. While viral contamination is certainly important, it is neither new nor is it peculiar to abnormal cells. The discovery of viruses in cell substrates and in products has occurred several times since the early days of polio vaccine. The first major viral contaminant was, of course, SV-40 virus in poliovirus vaccine produced from primary Rhesus monkey kidney cell cultures. In addition to endogenous viruses, there is also the possibility that an exogenous virus could be introduced into the cell culture at any point in its *in vitro* history. Regulations were therefore developed many years ago to test for known endogenous viruses in primary cells used in vaccine production and for a spectrum of exogenous viruses in diploid cell lines. The opportunity for the introduction of an exogenous virus in a continuous cell line is obviously greater than for a primary cell or diploid cell line, both of which have a limited in vitro life potential. The concept of a cell seed and the standardization of a frozen cell bank is probably the single most important feature which helps to assure the absence of both endogenous and exogenous viruses from cell substrates. However, from public health and safety points of view, there has been room for the use of cells which contain endogenous viruses. A specific example is the presence of avian leukosis viruses in chickens and in embryonated hen's eggs. Until recently, all yellow fever vaccines produced in the world used such eggs, and the vaccines contained live avian leukosis viruses. Because of the theoretical possibility that there may be long term adverse health implications to having administered live avian leukosis viruses to humans, a large retrospective study was undertaken, and the results showed no increased health risk for cancer in the recipients of the vaccine. The presence of the avian leukosis viruses therefore did not seem to be a safety issue. Nevertheless, at the same time a new yellow fever seed virus was being established, attempts were made to free it of avian leukosis viruses so that in the future yellow fever vaccines could be produced without avian leukosis in the starting material. That experience established an important principle which might be helpful in making decisions on products derived from abnormal cell lines. Namely, that the mere presence of a live viral contaminant in a final product was not sufficient to eliminate that product from the market. Instead, the benefits of the product, and its track record with regard to safety, were balanced against the theoretical risks of the avian leukosis viruses.

A variation on this same theme is also worth mentioning. In addition to yellow fever vaccine, fertilized hen's eggs are used to produce influenza vaccines. But our currently licensed influenza vaccines are inactivated, as opposed to the yellow fever vaccine which is live. In fact, such eggs were used to produce influenza vaccine many years before avian leukosis viruses were even discovered. That meant that there had already been a large amount of clinical experience with data showing that the vaccine is both safe and effective. No attempts were therefore made to eliminate the use of avian leukosis-containing hen's eggs in the commercial production of inactivated influenza vaccines anywhere in the world. And of course we all know that there have been no attempts at removing chickens or eggs from the food supply just because they both contain avian leukosis viruses. The reason, of course, is that in both situations the avian leukosis viruses are killed before they get to humans in the form of a vaccine, a cooked chicken, or scrambled eggs. The alternatives of using leukosis-free eggs for influenza vaccines or avian leukosis-free chickens for food, are not realistic options because of the difficulty and the expense involved in providing the numbers required. This example illustrates the point that even though a substrate may contain an endogenous retrovirus, the cell might still be considered acceptable depending on the specifics of the product and how it is manufactured. Whether the product is a biochemical or a replicating biological agent will make a difference. And whether or not a biological agent is live or inactivated also will make a difference. And, of course, data showing that prior clinical use of products derived from that cell system have been safe to use is of real importance, if it exists.

We have been trying to look at a whole range of factors in making decisions on clinical studies with new biotechnology products derived from abnormal mammalian cells. And I might just mention several examples to illustrate the major considerations. Interferon derived from human lymphoblastoid cells is probably the oldest example of major importance. That experimental product is produced in cells that originated from a human Burkitt's lymphoma. The cell line itself is tumorigenic and it contains part of the Epstein-Barr virus genome in its chromosomes. Two more recent cases are hybridomas to produce monoclonal antibodies for the detection of metastatic tumor cells, and the Chinese hamster ovary cell line to produce tissue plasminogen activator for the treatment of acute myocardial infarctions. In both of these last two cases the cells are again tumorigenic, and they might contain endogenous murine viruses. In making the decision to allow clinical studies to proceed, we took into account several major factors. First, none of the products were replicating biological biological agents, and there was therefore no possibility that any of the cell substrate DNA could be incorporated into the genome of the product. The second point was that the purification processes for the interferon, the monoclonal antibody, and the tissue plasminogen activator could each exclude significant amounts of RC-DNA, and there was reasonable assurance that live contaminating viruses were absent from the products. As with any experimental product, there were still theoretical risks, but they appeared to be heavily outweighed by the potential benefit to human subjects in proceeding with clinical studies.

The three major issues of residual cellular proteins, RC-DNA, and viral contaminants were discussed in detail at the recent workshop I mentioned previously, and there was a general consensus that the approach we have been taking to date in evaluating the acceptability of various types of abnormal cells in the manufacture of new drugs and biologicals is reasonable. Perhaps more importantly, there was general agreement that the dogmas of the past should not be used to hinder research and development of new products, and especially those derived from the new biotechnology. Rather, the acceptability of any given cell system should be based on the data available to support the safety of the product derived from from it. To facilitate a working dialogue with both industry and the academic community, we have prepared a series of documents describing tests which should be considered in the preclinical stages of recombinant DNA products, hybridomas, and more recently for cell substrates in general. Ultimately we hope to arrive at the point where there is enough consensus, and technological advances have slowed enough for these informal working documents to evolve into formal guidelines. But for the immediate future they will serve as at least a framework within which we can work and move towards decisions on specific products.

EXPERIENCE IN THE CULTIVATION OF MAMMALIAN CELLS ON THE 8000 l SCALE

A. W. Phillips
G. D. Ball
K. H. Fantes
N. B. Finter
M. D. Johnston

Wellcome Biotechnology Ltd.
United Kingdom

INTRODUCTION

During the last decade the production of biologicals has undergone a revolution as a result of the impact of genetic engineering. This has made commonplace the use of bacteria or yeasts for the production of polypeptide hormones, interferons and numerous other proteins. In view of these successful applications of the new technology, one may be tempted to ask why mammalian cells still require to be cultivated on a large scale. The answer is that, apart from the fact that it is not always possible to produce suitably engineered prokaryotic cells, there are special circumstances where the use of such cells does not provide the best solution.

At Wellcome, we have cultivated mammalian cells on a large scale for a number of purposes, as discussed below and we still think that the technology has much to offer.

A. **Vaccine Production**

The protective antigens of most viruses are so large, complex or variable in structure that it is not possible at present to produce them by procedures involving recombinant DNA. Thus, the traditional technology involving the production of the appropriate viruses in large scale tissue cultures remains the most usual approach in vaccine manufacture. An example of this in which we have been concerned for many years is the bulk manufacture of FMD vaccine in baby hamster kidney cells.

B. Production of Proteins Synthesized by Cells as a Result of Manipulation

The large-scale culture of manipulated eukaryotic cells has two main areas of application. In the first, the cells used contain engineered DNA and provide a single gene product. The application is of value when the required product is a glycosylated protein which cannot be obtained in adequate amounts from nonengineered cells. It is often possible to insert multiple copies of the gene concerned into a cell which can be readily cultivated. An example of this is the introduction of the gene of human β-interferon into hamster cells.

The second area is the cultivation of hybridoma cells for the production of monoclonal antibodies. These are cells which have been manipulated in a very specific way for a specific purpose. It seems likely in view of the infinite variety of antibodies which might be required and of their size and structural complexity, that for the foreseeable future they will continue to be produced in eukaryotic cells.

C. Production in Suspension Culture by Unmanipulated Mammalian Cells

The approach is only possible when cells producing the required proteins in adequate amounts can be obtained from natural sources and established as a continuously growing cell line. It is particularly advantageous when it is important that the protein produced should be in its natural glycosylated form or when the cells make a family of related proteins all of which are required. The latter situation applies to the α-interferons, which we have opted to manufacture by a method yielding a natural mixture of subtypes rather than by recombinant DNA technology which from a given clone should provide only one subtype.

In the next section we shall discuss the main problems we have encountered in establishing the production of the α-interferons on a commercial scale.

D. The Bulk Production of Human Lymphoblastoid Interferon

The background to our decision taken ten years ago, to produce interferon from suspension cell cultures, is that at the time the annual supply of human interferon totalled only about 0.5 g or some 20,000 human doses. This was virtually all prepared from leucocytes derived from transfusion blood by Dr. K. Cantell and his colleagues. Much more criteria was required for clinical trials and we considered possible alternative sources.

At that time we had some ten years experience in growing baby hamster kidney cells in suspension in 1000 l tanks for the production of foot and mouth disease vaccine. In view of this background experience, we were understandably attracted to the prospect that some cell cultured in very large tanks might yield enough interferon to make possible the clinical studies which were so clearly required.

The first requirement was to identify a suitable cell.

1. The Cell

The need was to find a cell which could be cultivated continuously and be considered suitable as a source of α-interferons for use in man as a therapeutic agent.

In 1975, Strander[1] and his collaborators had described the production of interferon from a number of continuous lines of B-lymphoblasts after treatment with Sendai virus. The most productive was the cell line called Namalwa, derived from a child of that name who had a Burkett lymphoma. We screened more than 100 different lymphoblastoid cell lines for their ability to produce interferon[2] and at the end it still seemed that Namalwa was the best. Although it was far from clear at that time that licensing authorities would accept a product generated by these cells because of their origin, it was agreed recently at a workshop organized by the F&DA, that with appropriate safeguards, transformed cells can be used for the manufacture of therapeutic agents. Fortunately it has proved possible to devise a procedure for the purification of interferon which eliminates every one of a variety of infectious and transfecting agents when added as marker substances to the

---

[1,2] see References

crude harvest.[3] With these and other data we have been able to gain the necessary approval to carry out a large number of clinical trials in many countries.

As a necessary first step in the production process we grew a substantial quantity of cells, dispensed them into vials and stored them in liquid nitrogen as a master cell bank. Cells from the bank were grown and shown to be free from extraneous agents. To start production a vial from the master cell bank is thawed and the revived cells are cultivated as spinner cultures of increasing size until there are enough to seed the smallest production tank.

At intervals a new vial of cells is taken from the bank and revived so that the entire population of cells in use in production can be replaced.

2. The Growth Medium

We have found that the Roswell Park Memorial Institute medium 1640[4], developed for the cultivation of lymphoblasts, is suitable for Namalwa cells. We prepare concentrates of the constituent salts, amino acids and vitamins, store these frozen and use them to manufacture the medium when required.

Serum is traditionally a constituent of tissue culture medium and foetal serum or the serum of newborn animals is widely regarded as being the most suitable. Quite apart from the cost of such serum, it would be difficult to obtain it in amounts sufficient for our scale of production. Fortunately we have found that adult bovine serum is as effective as the serum of younger animals and we now use it exclusively. The risk of contaminating cultures with bovine viruses or mycoplasmas by the use of this serum is eliminated by exposing it to γ-irradiation. The medium used for growing Namalwa cells contains 3.5% of this serum.

As tissue culture media contain heat-sensitive ingredients, they require to be sterilized by filtration. We use disposable filter cartridges to achieve first clarification and finally sterilization to 0.2μ.

---

[3],[4] see References

3. Conditions of Culture

Temperature and pH require to be maintained near to physiological values. To provide qood growth and high viability, it is also necessary to aerate the cultures and provide agitation.

Under optimized conditions, the doubling time for Namalwa cells averages about 48 hours with maximum densities of viable cells of the order 4-5 x 10^6/ml.

4. Outline Method of Production

The following are the steps needed in the large scale manufacture of Namalwa interferon:

1. Establish and maintain large breeder cultures of cells.
2. At intervals transfer cells to production vessel and add sodium butyrate (1-5 mM).
3. About 48 hours later add Sendai virus to production vessel.
4. About 24 hours later remove the cells and harvest the crude interferon.

Butyrate is added in step 2 because Namalwa cells otherwise yield variable amounts of interferon on different occasions when induced with Sendai virus.[5] To be a satisfactory inducer of interferon, the Sendai virus used must contain infective virus and defective interfering particles in optimal proportions.[6] Our manufacturing method for Sendai virus is designed to achieve this mixture.

5. The Plant

As a minimum requirement, the production plant must provide the following facilities:

1. Media preparation and filtration.
2. Breeder vessel.
3. Backup vessel(s).
4. Production vessel.
5. Separation equipment for cells.
6. Harvest vessel.

[5,6] see References

The function of the breeder vessel and the production vessel have already been mentioned: the backup vessels which are of smaller capacity provide cells for reseeding the breeder vessel in the event of loss.

As a result of our long experience in this general area, all our development and production is carried out in fermentation vessels of our own design. Even when our first pilot plant for interferon production was built it was of such a size that by the standards of the time it could appropriately have been termed a production plant. It employed a 600 l breeder vessel and a production vessel of 1000 l. In this plant we learned much about the design of suitable equipment for interferon production. For example we confirmed the value of our system for indirectly coupling the drive from the stirrer motor to the shaft of the impeller by the use of large ceramic magnets so allowing stirring in a completely sealed tank. On the other hand, the removal of the cells from the crude interferon at harvest by passage through a bed of diatomaceous earth though simple, proved cumbersome and slow. In our second plant, the production vessels had a capacity of 3000 l and the breeders of 2000 l. At harvesting, cells were removed either by filtration on disposable units or by continuous centrifugation. Both these methods for removing cells work very well though each has its disadvantages. Thus filtration has a high operating cost in terms of throwaway filters while the centrifuge has a high capital cost and cleaning after use presents problems. This plant which has been in use for just over four years, has produced interferon equivalent to $4 - 8 \times 10^6$ doses: enough for a major clinical trial programme. Our largest plant uses vessels with a capacity of 8000 l with a correspondingly increased output.

Throughout the development of these plants a major technical objective has been to avoid bacterial contamination of the breeder vessels while making a large number of additions to and withdrawals from them each week. In our first plant the mean life of breeder culture was only about one month, in our second plant it was nearly three months, while in our latest plant a life of nearly six months is usual. Under these circumstances the loss of a breeder vessel is more often due to mechanical or instrumental failures than to contamination by bacteria. This improvement has been achieved by paying meticulous attention to detail in both the design and operation of the plant rather than in the adoption of any revolutionary concept in the design.

The performance of the large plant has been very gratifying, probably as a result of experience gained in operating the older plants. Thus the cells reach higher maximum counts and yield more interferon than in our earlier and smaller plants.

We believe that the mass cultivation of cells in suspension will have a role in biological manufacture for some years to come. In systems other than that described the problems arising are likely to be similar to those we have encountered. With appropriate technical resources and experience with large-scale fermenters, all of these can normally be overcome.

ACKNOWLEDGEMENTS

The authors acknowledge with gratitude the contribution made to this work by all the members of the Interferon Project, in particular by the three plant managers, Mr. K. F. Pullen, Sr., J. Estefanell and Mr. C. J. Burman.

REFERENCES

1. Strander, H., Mogensen, K.E. and Cantell, K., J. Clin. Microbiol. 1, 116 (1975).
2. Christofinis, G.J. and Finter, N.B., Proc. Symp. Interferon, Yugoslav. Acad. Sci. Arts, Zagreb,p.39 (1977).
3. Finter, N.B. and Fantes, K.H. in "Interferon" (I.Gresser, ed.),p.65. Academic Press, London 1980.
4. Moore, G.E. Gerner, R.E., and Franklin, H.A., J. Am. Med. Assoc. 199, 519 (1968).
5. Johnston, M.D., J.Gen.Virol. 50, 191 (1980).
6. Johnston, M.D., J.Gen.Virol. 56, 175 (1981).

DISCUSSION OF THE PAPER

DR. R. G. RUPP (Damon Biotech, Inc., Needham Heights, MA): I have two questions. First, what's the quality of the water that you are using to make your medium?

DR. A. W. PHILLIPS: This is water which we make from our main's water by deionizing and reverse osmosis. It's filter sterilized and stored sterile in a large tank piped around to all the tanks.

DR. R. G. RUPP (Damon Biotech, Inc., Needham Heights, MA): The 8,000 liter tank or any of those large culture vessels require quite a bit of medium just to maintain the culture. Do you have a holding facility someplace for your medium; do you have large tanks of medium sitting around someplace in a cold room?

DR. A. W. PHILLIPS: We have the concentrates of the various constituents the salts, the vitamins, the amino acids and so on, and when we need to make medium, say on a Monday, which in fact is very common, we will knock out the required quantity and filter it into the vessels.

DR. R. G. RUPP (Damon Biotech, Inc., Needham Heights, MA): So you don't QC or do a quality control of each batch?

DR. A. W. PHILLIPS: Not the blended, final medium, no we do not.

DR. J. FEDER (Monsanto Company, St. Louis, MO): Do you hold the medium batch to test for sterility?

DR. A. W. PHILLIPS: No. We take a sample from the vessel every time we carry out a manipulation. Although, of course, if the medium is contaminated, we tend to know before we get the results of the test.

DR. J. FEDER (Monsanto Company, St. Louis, MO): Are there antibiotics in the medium?

DR. A. W. PHILLIPS: Yes, there are.

DR. R. W. ELTZ (Monsanto Company, St. Louis, MO): I don't know if you'll share with us your contamination record of viral contamination if you have any, but I wonder if you might share the design features in both air preparation for your reactors and medium filtration that mitigate against viral transmission into the system?

DR. A. W. PHILLIPS: Viral transmission is something which we certainly don't see in terms of cytopathic effects observable within the tanks themselves. That is something we have never seen. There is no specific treatment to remove virus from the medium or from the air. The safeguard which we have is in the later processing of the interferon where interferon is such a very stable protein that we have been able to use quite vigorous reagents in its purification and as I said earlier on, we have been able to demonstrate the elimination of all of the effective agents that we'd used as a challenge. So the answer to the question really is that we have no evidence of virus in the cultures but that we would not regard it as a source of grave concern if there were a trace of those.

DR. R. W. ELTZ (Monsanto Company, St. Louis, MO): Could you amplify the nature of the filtration system of the medium itself?

DR. A. W. PHILLIPS: It is filtered through cartridges. The final filtration is done with nylon cartridges supplied by Paul.

DR. D. MUELLER (Hoffmann-La Roche, Nutley, NJ): I'm interested in your method of agitation. You said you had a magnet encased in ceramics. Now is that a bottom drive or is it a top drive?

DR. A. W. PHILLIPS: That's a top drive.

DR. D. MUELLER (Hoffmann-La Roche, Nutley, NJ): Do you have conventional turbine paddles on there?

DR. A. W. PHILLIPS: Yes, it's the sort of thing you might find on the back end of a boat, I think.

CALLIN CAIN (Drexel University, Philadelphia, PA): Regarding the redox potential as one of the parameters, what kind of effect do you really see and how do you maintain the redox potential in 8000 liter fermenter? What effects do you see in terms of production of interferon and what is the level you maintain?

DR. A. W. PHILLIPS: That is maintained at a level which is pretty highly oxygenated and as such enables the cells to survive and grow quite happily. I think that with the type of standards that one has for that sort of measurement, it's probably the sort of thing that's not easily transferable from one sort of equipment to another. If I give you a millivolt reading of around 300 millivolts related to a calomel half cell, you may very well find that it wouldn't work too well in your system.

DR. J. FEDER (Monsanto Company, St. Louis, MO): Dr. Phillips, we've heard some comments about high density cultures earlier in the day. There is, conceptually, the possibility of having densities that exceed tenfold that which one finds in a static reactor such that one would have $3x10^7$ cells/ml instead of $3x10^6$ cells/ml. Do you have any feelings or any thoughts about that versus what one would consider very impressive conventional large-scale culture?

DR. A. W. PHILLIPS: Yes, I mean it's a great thing! When I've seen it work, I'll tell you what I think. Conceptually, it is a great idea, but I doubt if we'll go away and scrap our big vessels.

PERFUSION CULTURE SYSTEMS FOR PRODUCTION OF MAMMALIAN CELL BIOMOLECULES

William R. Tolbert
C. Lewis, Jr.
P. J. White
Joseph Feder

Monsanto Company
St. Louis, MO

During the past few years the need to obtain products from large-scale _in vitro_ culture of animal cells has greatly accelerated. While viral vaccines had been the traditional commercial cell culture product, interferon and monoclonal antibodies have caused a major reevaluation as evidenced by today's symposium. On the horizon are many more products with potentially very high value human therapeutic or _in vivo_ diagnostic applications which will most efficiently be produced from genetically modified animal cells grown _in vitro_, such as human tissue type plasminogen activator and blood factor VIII.

Monsanto Company has developed over the last 12 years a broad based, new cell culture technology which is now most appropriate to current enhanced production needs in this area. This proprietary technology differs significantly in conception through practical application from the more conventional approaches to large-scale growth of both anchorage-dependent and independent cell lines and for products secreted into the medium, induced by special procedures, or retained within the cells. Over this time more than a quarter million liters of rodent and human cell suspensions have been grown and harvested. Development of this new technology was based on the principle that new types of equipment should be designed to efficiently meet the requirements of antibiotic-free vertebrate cell culture rather than modification of conventional fermentation equipment. This required development of systems which would maintain absolute sterility to protect the cells with proliferation rates measured in hours from contaminants with proliferation rates measured in minutes and also required an extremely high confidence level for each operation to prevent loss of the expensive serum supplemented medium. The systems developed to meet these demands required that equipment be relatively simple, portable and autoclavable, and that the facility be designed with increasing levels of containment both both for protection of personnel and isolation of cultures. The ability to maintain high flexibility was also extremely important as each cell type and individual cell line has its own specific requirements and idiosyncrasies for optimum growth and product yield.

The first major component developed for large scale culture was a 100-liter vibromixer agitated reactor that was portable, autoclavable and relatively simple in construction. (See Figure 1.) This reactor has been operated in batch and semi-batch modes for production of hundreds of thousands of liters of suspension culture and hundreds of kilograms of wet packed cells. It is useful for cells that grow freely in suspension or as aggregates or small clumps of cells in suspension. Both cells and expended medium can be obtained as products usually harvested by centrifugation. Due to the higher shear fields in the region of the vibrating plate agitator, it is not recommended for cells attached to microcarriers. It is particularly useful, for this same reason, with cells that tend to form excessively large clumps. These large clumps inhibit diffusion of oxygen and nutrients causing necrotic centers, and the vibromixer action tends to reduce aggregate size to allow better growth response.

While this and other batch growth systems have been successfully used for many cell types over long time periods, they result in basically a feast and famine environment for the cells. The cells are initially exposed after inoculation or feeding with high levels of nutrients and with lower levels of waste products. As the cells grow, and utilize these nutrients, they produce waste products such as lactic acid from glucose and ammonia from amino acid metabolism. Under these conditions most cultured cells begin to reduce their growth and die at a level of about 1 to 2 million cells per ml or one ml of wet packed cells per liter of suspension. Figured at 1,000 ml of packed cells per liter of tissue, this cell death starts at one one-thousandth of tissue density. That is, conventional culture techniques, even with the high degree of selection necessary for <u>in vitro</u> growth, is at least 1,000 times less efficient than the <u>in vivo</u> environment.

In nature the mammalian cell is part of an organism whose structure assures a homeostatic environment in which the supply

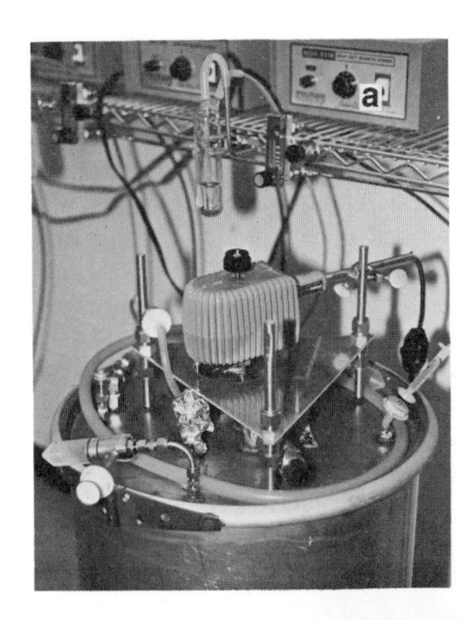

Fig. 1. 100-liter vibromixer agitated culture reactor is shown in warm room during growth state (a) and during harvest with Western States STM1000 centrifuge (b). Agitator disks and other internal parts are shown in (c). Reprinted from Tolbert et al. (4).

of nutrients and regulatory factors are maintained and metabolic wastes removed. The need to perfuse systems to improve nutrient supply, remove metabolic products and stabilize the *in vitro* environment was recognized early in the history of tissue culture. According to Paul Kruse, Jr. (1), the credit is due to Burroughs as far back as 1912 for the first attempts to furnish cultures of animal cells with a continuous supply of fresh nutrient medium. Attempts to develop such systems were made early by de Hahn in the 1920's and 1930's, and Pomerat and Rose in the 1950's and 1960's (1). The use of perfusion systems to achieve a steady-state environment, in fact, did provide significant advantage in most of these studies over static growth methods. In late 1960, Himmelfarb, Thayer and Martin (2) developed a spin filter device for perfusing a suspension culture; they demonstrated they could achieve much higher cell densities than previously obtained by conventional suspension systems.

This paper reviews the development of several large-scale mammalian cell perfusion systems for the growth of both anchorage-dependent and independent cells. For additional information, the reader is referred to an earlier report (3). This paper also describes an entirely new perfusion system for maintenance of cells at high densities in essentially non-proliferative status. The development of an efficient mammalian cell perfused suspension system which could achieve both high cell density growth and chemostat-like operation was achieved in stages. The system first developed consists of a 4, 12 or 40 liter glass culture vessel containing a rapidly rotating, vertically disposed porcelain filter for removal of suspended medium (Fig. 2).

This filter had a pore size of 1-2 microns and was rotated at 200-300 rpm to prevent clogging of the filter pores and to provide agitation to the cell suspension. While this rotating filter system was effective in growing cells to densities as high as 3×10^{7} cells/mℓ it had some deficiencies for long term production. Figure 3 shows a diagram of the perfusion chemostat system. This represents the next stage of development of our perfusion systems. In this new system, a stationary filter surrounded by a rotating concentric agitator substituted for the rotating filter. This greatly simplified the system by removing the need for a rotating seal assembly, making all liquid seals static. The clogging of the filter pores was prevented by tangential shear caused by the concentric agitator-induced motion of the suspension around the stationary cylindrical filter. A second important improvement was to remove the filter assembly entirely from the growth vessel and locate it in a separate satellite vessel (Fig. 4). This satellite vessel was connected in a recycle circuit to the main growth vessel with suspension pumped back and forth to maintain equivalent component concentrations. Expended medium, containing product in most cases, was withdrawn from the filter in the satellite vessel and replaced with fresh nutrient medium in the growth vessel. This vessel could be operated in a regular growth mode or a chemostat mode of operation. The chemostat mode of operation was implemented by pumping both cells and expended medium from the growth chamber to a separate holding vessel maintained at 4°C. A third improvement in this system involved the use of a novel, extremely gentle, agitation system

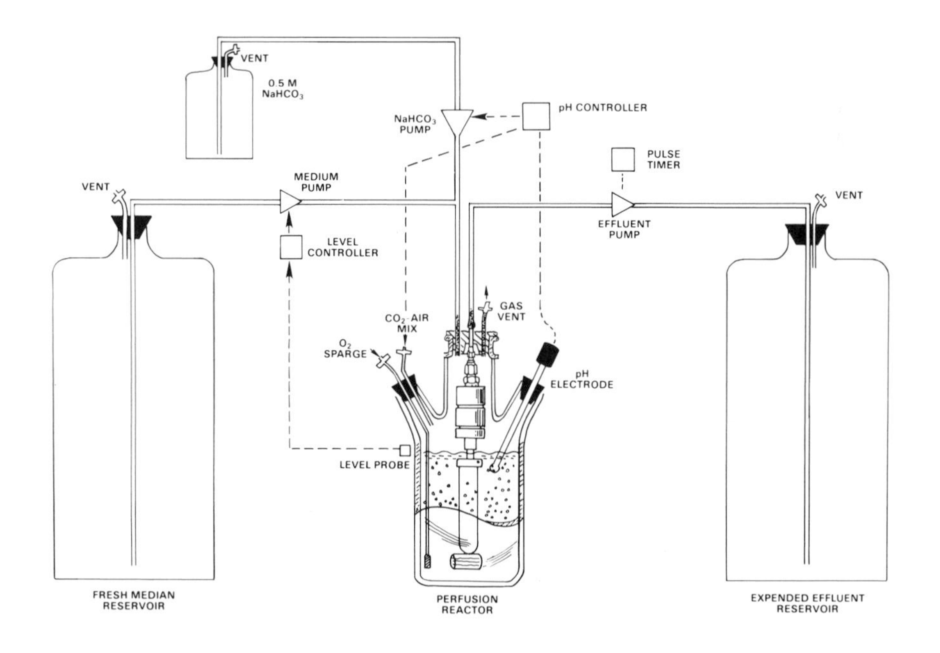

Fig. 2. High density cell suspension is grown in the central reactor with agitation by 200 to 300 rpm rotation of the vertical filter. Cell-free expended medium is withdrawn through this filter to the effluent reservoir at a rate set by the peristaltic pump/timer combination. The capacitive level control system maintains the reactor volume by initiating fresh medium flow to replace the effluent. pH is monitored by an autoclavable INGOLD® electrode and controlled both by varying the CO_2 concentration of the overlay gas and by addition of 0.5 M $NaHCO_3$ when required. Dissolved oxygen concentration is maintained by a very low volume 100% O_2 sparge through sintered glass. All gases enter vessels through 0.2µ membrane filters. Reprinted from Tolbert, Et al.(11).

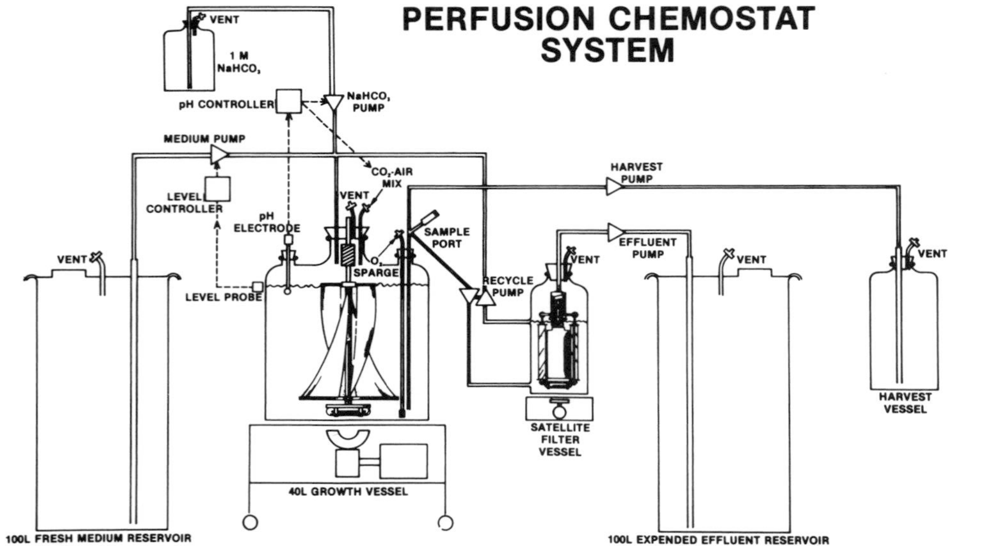

Fig. 3. A diagram of the entire perfusion chemostat system is shown. Cell-free expended medium was removed from the interior of the porcelain filter at a rate set on the effluent pump to provide the required perfusion of fresh medium. Cell suspension was recycled to and from the satellite vessel by a double-headed pump at a rate sufficient to maintain similar component concentrations in the two vessels. While operated as a chemostat, a harvest pump removed cell suspension directly to a harvest vessel contained in an adjacent 4°C cold room. The resulting reduction of the liquid level in the growth vessel was compensated by an exterior capacitive level probe which actuated the medium pump to supply fresh nutrient medium. An autoclavable pH electrode penetrating the growth vessel allowed control of the pH by varying the CO_2 in the mixture of overlay gas and by the addition of sodium bicarbonate-sodium hydroxide mixture when necessary. A sterile air-shielded sample port and a means for low rate oxygen sparge through a sintered glass dispersion tube were also provided. The growth vessel was located in at 37°C warm room. Reprinted from Tolbert and Feder (3).

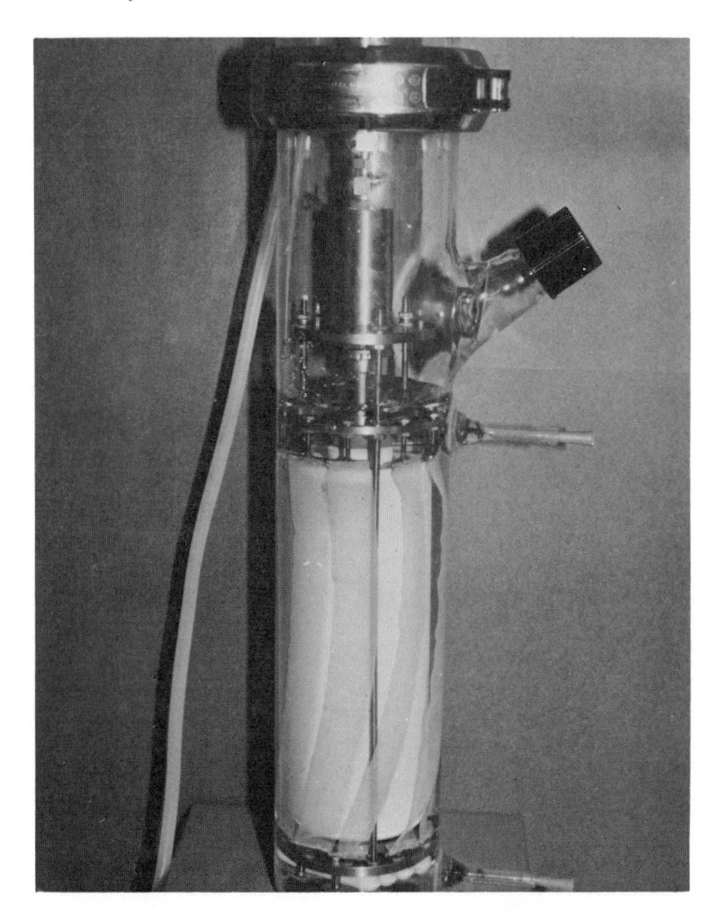

Fig. 4. Satellite filter vessel with 1 μ porosity, stationary porcelain filter surrounded by a concentric agitator. This agitator, rotating at 40 to 80 rpm, utilizes nine flexible "sails" to prevent clogging of filter pores. See Fig. 3 for description of operation.

utilizing large, slowly rotating, flexible sheets or sails. By increasing the surface area of the agitator, the energy transferred per surface area is significantly reduced. These sail agitators were operated at as low as 10 rpm to provide the required mixing with greatly reduced stress to the high density, fragile cells. This also was employed in a different arrangement as the concentric, rotating agitator surrounding the stationary filter. In the growth vessel four large, flexible sheets were used; in the satellite filter vessel the concentric agitator consisted of nine or more flexible sheets running the length of the stationary porcelain filter (5). These were magnetically driven at a rotation of about 50-75 rpm which was sufficient to prevent filter clogging during removal of hundreds of liters of expended medium. The satellite vessel also was designed for easy replacement, if necessary, without requiring termination of the run. These perfusion systems allowed growth of mammalian cells at 10-15 ml of wet packed cells per liter or approximately one hundredth of tissue density.

In addition to these improvements of the perfusion chemostat, an aseptic connection system was devised (6). Operation of the perfusion chemostat or other large systems, particularly in the absence of antibiotics, was facilitated by the ability to make aseptic connections in non-sterile environment. This aseptic connection system consisted of commercially available quick-connects such as SWAGELOK QC series connectors surrounded by a cylindrical jacket. As shown in Figure 5, this jacket has a side port for injection of sterile air which flows around the connector and is forced out through a narrow gap between the end of the connector and the flanged end of the jacket creating a conical barrier of sterile air between the external environment and the ends of each connector. This barrier of sterile air illustrated by water in the figure prevents contamination from entering the system during the connection, or disconnection operations. When not in use the connectors are protected by sterile plastic shields. A modified version of this device is used to allow repeated sampling of the reactor. It is particularly appropriate where cell aggregates or microcarriers are present which would prevent successful use of a syringe-needle, septum system.

Another development used with the perfusion chemostat system is a gentle, pumping device particularly useful for transfer of biological fluids containing fragile components (see Figure 6). This system consists of a length of collapsible and flexible tubing contained within a fluid filled chamber. This chamber is connected to a piston-cylinder arrangement. When the piston is withdrawn, the flexible tube is expanded, and when the piston returns, the tubing collapses. At either end the tubing is connected to gravity actuated check valves so that the pulsing or breathing motion of the flexible tubing causes liquid to be moved up through the pump. Since the check valves are designed to be symmetric, when the pump is inverted, the flow of liquid is reversed in the external circuit. This allows filling and emptying of the filter vessel without changing connections. Also, since there is no frictional insult to the tubing compared to peristaltic pumps, tubing life is essentially unlimited. If any leak did occur

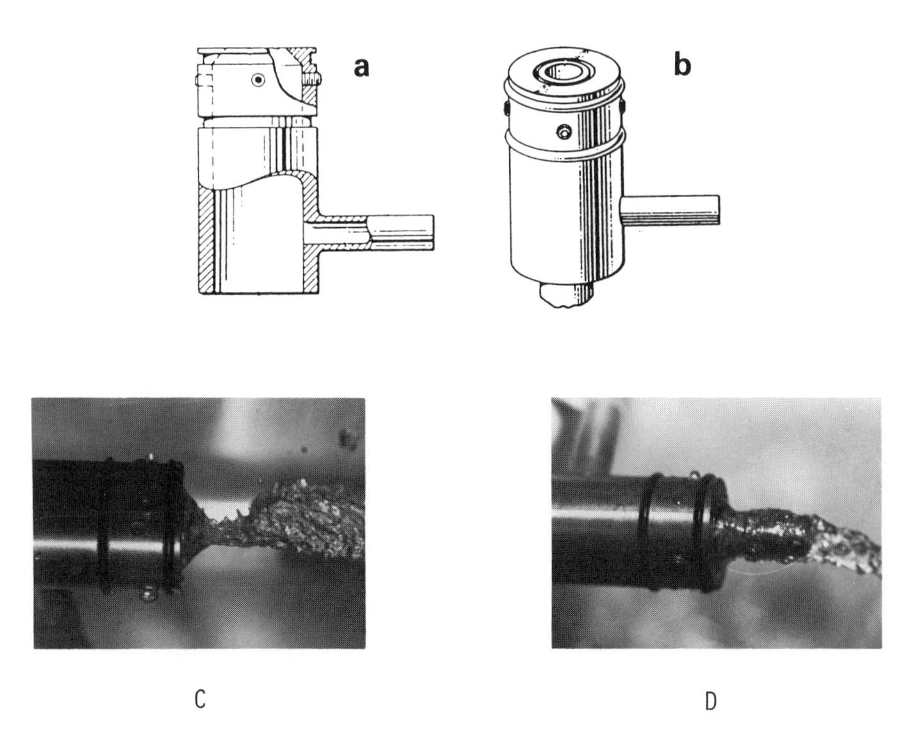

Fig. 5. The cylindrical jacket for the air-shielded connectors is shown (a) in partial cross-section illustrating the 45° internal flange which forms the conical air barrier and (b) with a female quick-connect inserted and O-rings in place. Both female (c) and male (d) shielded connectors are shown with water in place of sterile air to illustrate formation of the protective cone over the ends of the couplers. Reprinted from Tolbert et al. (6)

in this system, the external pumping system would prevent any possible biohazardous spill. These pumps are used particularly between the growth and filter vessels of the perfusion chemostat system where fragil cells and cell aggregates at high density must be circulated.

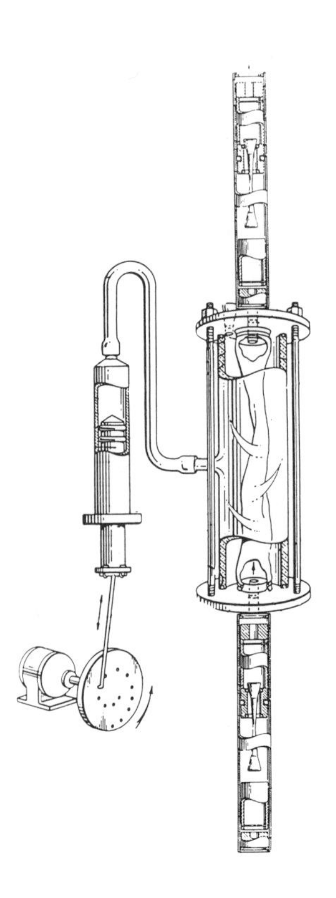

Fig. 6. Gentle pumping device to transfer single cell and cell aggregate suspensions between perfusion reactor and satellite filter vessel. See text.

Figure 7 shows a typical growth curve for the rat Walker carcinoma-sarcoma cells grown in suspension in a conventional spinner. A doubling time of about 16 hours was observed and after some 50 hours, a maximum cell density of 1-2x10^6 cells/ml was obtained. The cell viability began to fall at this time. An examination of the glucose metabolism revealed an almost stoichiometric conversion of glucose to lactate. Similar results have been observed with numerous types of cells grown both in suspension and in monolayer in T-flasks. In the absence of perfusion, particularly where the oxygen was somewhat limiting, one obtains a stoichiometric conversion of glucose to lactate. There is no intention to suggest that lactate _per se_ is toxic to cells. If the pH is maintained, then lactate in itself has not been found to have untoward effects to the cell

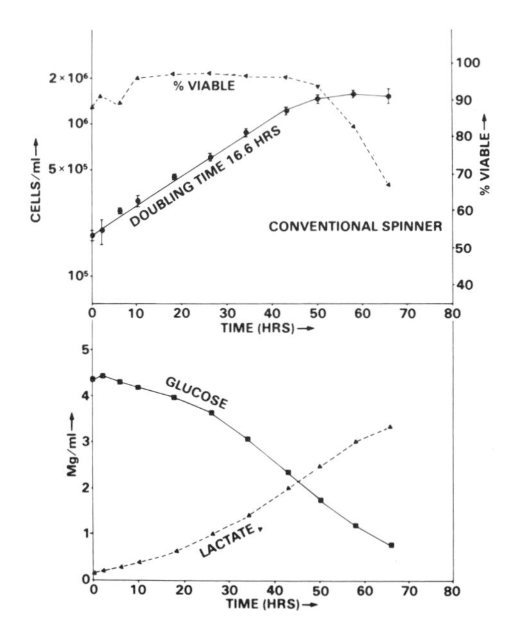

Fig. 7. Rat carcinosarcoma suspension was cultured in a conventional Bellco spinner. Cell density determined by hemocytometer ± standard deviation of replicate counts. Viability is measured by trypan blue dye exclusion. Cells were pelleted by centrifugation for 10 min at 300 x g and supernatant solutions frozen for later glucose and lactic acid determination. Reprinted from Tolbert et al (4).

growth. The glucose-lactate data exemplify the accumulation of a metabolic product in the medium in the absence of perfusion. However, together with the glucose metabolism, there are other products which accumulate in the medium in a similar fashion which often prove to be toxic. An example is the accumulation of ammonia seen in many cultures as reported by McLimans (7).

Figure 8 presents a growth curve and glucose and lactate concentrations in medium during the culture of these same cells in the 4-liter rotating filter perfusion reactor. A similar doubling time was observed; however, cell densities of $3x10^7$ cells/ml were achieved, an almost thirty-fold increase over that possible in the conventional suspension reactor. The viability was maintained at near 100% and the levels of glucose and lactate remained at a near steady state concentration. A comparison of the cell yield obtained in the perfusion reactor with that of 100-liter conventional batch system shows that the perfusion reactor produced 1.7×10^{12} cells, using 621 liters of medium, while the 100-liter batch system produced 13.4×10^{12} of these same cells using 14,000 liters of medium. The medium efficiency, therefore, was 2.7-fold higher in the perfused system.

Figure 9 - showing an experiment to test serum requirements in which the SK-HEP-1 human hepatoma cells were grown in the 12-liter perfusion reactor, run in the chemostat mode. Initially the cells were inoculated at 5% bovine calf serum and they exhibited a population doubling time of 57 hours; on day 7 the medium was changed to 2.5% bovine calf serum. The chemostat mode was initiated on day 12. The chemostat operation was adjusted to hold a cell density in the reactor of approximately 10 mls of packed cells per liter of medium and this was maintained through day 48 when the reactor was terminated. At day 21, the perfusion medium was changed to a 1.5% bovine calf serum and at day 37 to 1.0% bovine calf serum. The cell yield in each of these periods was 2.27 mls per liter for the initial growth period, 1.19 in 2.5% bovine calf serum at the beginning of the chemostat mode, 1.47 mls per liter in 1.5% bovine calf serum and 1.17 mls per liter in 1% bovine calf serum. If one

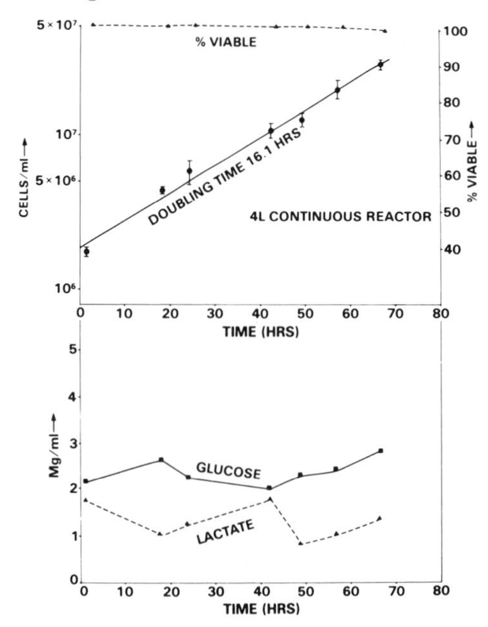

Fig. 8. Rat carcinosarcoma suspension was cultured in the 4-liter rotating filter perfusion reactor as described in the text. Cell density, viability, glucose and lactic acid determinations were made as described under Fig. 7. Reprinted from Tolbert et al (4).

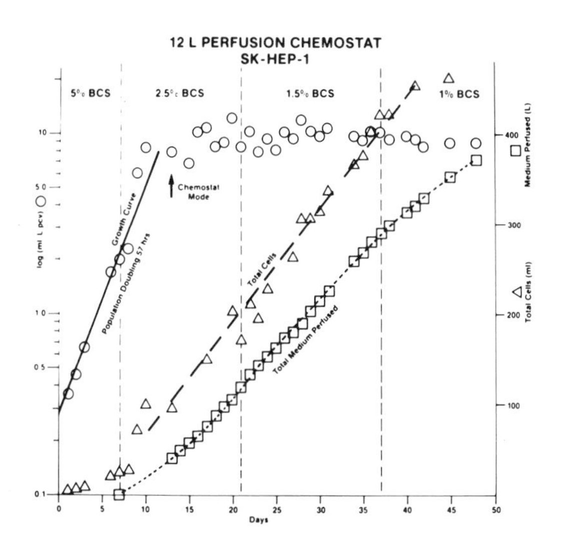

Fig. 9. Growth curve for human hepatoma cells grown initially in 5% bovine calf serum supplemented DMEM without antibiotics. Serum concentration was reduced as indicated with chemostat operation initiated on day 10 to maintain level of cell density.

normalizes the yields to the liters of serum in the medium, then one sees that at 5% bovine calf serum this is approximately 45.3; at 1.5% bovine calf serum, almost 117 mls of packed cells were being obtained per liter of calf serum used. Thus it appeared that the steady-state level of serum required in a perfused culture system was significantly lower than that required in a conventional static batch reactor, and a significant increase in the cell yield per liter of medium/serum used for growth was obtained with perfusion. One is tempted to suggest that there are components in the bovine serum which are depleted by the cells during growth. One must maintain a steady-state saturating concentration of these components which is lower than the 5% or 10% bovine calf or fetal calf serum which is generally used. With perfusion, a much lower steady-state concentration needs to be maintained than in the conventional batch reactor. While serum is a major supply cost for large scale production, it has an even more drastic effect on the downstream processing needed to purify products normally in concentrations of nanograms to micrograms per mℓ from the high milligram per mℓ concentrations of serum proteins. The greatest economic benefit in reducing serum levels is in this area.

These perfused suspension culture systems have been used to produce a variety of biomolecules including monoclonal antibodies. Table I summarizes the production of three different monoclonal antibodies using these systems. As an example, a total of 600 liters of the 9.2.27 hybridoma conditioned medium was produced in a 40 liter perfusion reactor which yielded 40 grams of monoclonal antibody. A minimum fetal bovine serum concentration of 1.0% was used to culture the cells. The average residence time for the medium in the reactor was 1.2 days with an average antibody level of about 67 mg/liter.

To this point, the comments on perfusion culture have addressed primarily suspension cultures. A perfusion system based upon the systems that have been described for suspension grown cells has been developed for growth of anchorage-dependent cells attached to microcarriers. The use of microcarriers is probably the most widespread system for industrial application. It first was developed by van Wezel in 1967 (8). In this system, small solid or semi-solid beads from one hundred to several hundred microns in diameter are suspended in nutrient medium and anchorage-dependent cells harvested from other growth systems allowed to attach to these beads. The cells then can grow to cover the beads under conditions of very gentle agitation in a conventional submerged culture system. The initial beads used by van Wezel as microcarriers were DEAE-Sephadex A50 ion exchange chromatography beads (8).

Fifteen years after van Wezel's original paper, the microcarrier method for growth of anchorage-dependent cells still represents the only system which has had significant large-scale application at an industrial level. Vessels in the range of hundreds of liters have been used for the production of vaccines and a scaleup of an order-of-magnitude larger is possible. The microcarrier system, however, with its advantages still has some basic difficulties in applications of many types of anchorage-dependent cells and the need for further development of specific equipment to exploit its potential motivated development of the system to be described. Some of the problems involve the scaleup of smaller to larger microcarrier cultures.

A diagram of the microcarrier reactor system is shown in Fig. 10. It is based upon the perfusion suspension culture system with a satellite filter vessel. It differs from that system by the addition of a settling bottle in the recycled circuit between growth and satellite filter vessels. The large surface area flexible sheet sail agitators described earlier for the perfusion chemostat allow successful operation with

Table I. Monoclonal antibody production by perfusion culture

Hybridoma (antigen)	Reactor volume (liters)	Culture time days	Serum Care minimum (per cost)	Cell Density maximum (ml/liter)	Total media (liters)	Total antibody (grams)
F11-A1-Bb (Methionine Bovine Somatatropin)	12	11	2.0	11.2	84	2.5
16-1-2N (H-2, KK, DK)	12	10	6.0	12.0	48	7.0
9.2.27 (M14 Human Melanoma Tumor)	40	18	0.5	8.4	600	40.00

much higher microcarrier concentrations than previously possible. Agitation of these reactors was in the range of 8-15 rpm and concentrations as high as ten grams per liter of biocarriers (Bio-Rad Laboratories) were successfully used. Carriers were sterilized by autoclaving at 124°C for 60 minutes in pH 5 buffer containing 0.05 M NaCl and 0.05 M MES [2-(n-morpholino) ethane sulfonic acid]. During autoclaving the depth of settled beads was less than two inches. After cooling the beads were aseptically transferred to the microcarrier reactor and washed twice with phosphate buffered saline and once with serum-containing medium. At this bead concentration, approximately one-fourth of the operating volume of the reactor consisted of settled beads. The 4-liter reactor containing between 30 to 50 grams of beads was inoculated with human diploid foreskin fibroblasts AG1523 cell line grown in fifteen to twenty 690 cm^2 glass roller bottles. Cells attach rapidly to the beads and spread out with 60-80% of the beads containing one to three cells with a volume density of between 1 and 3×10^5 cells per ml. Constant agitation was maintained during inoculation, cell attachment and through cell growth. Cell numbers were determined by modification of the nuclei counting procedure reported

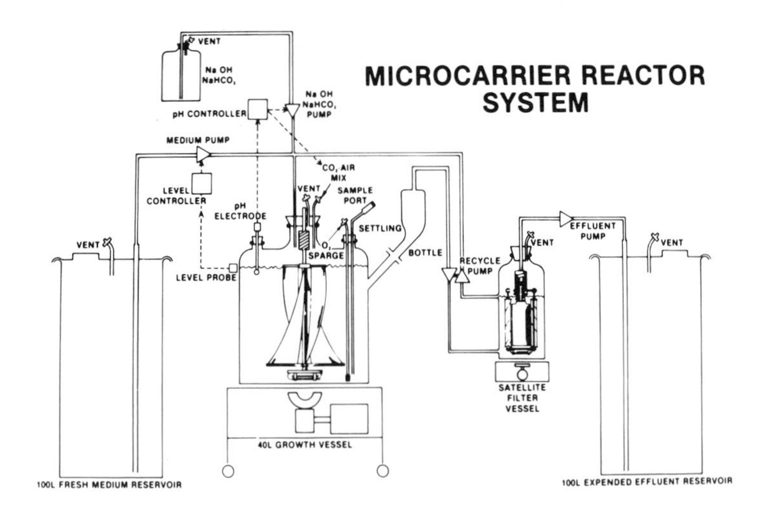

Fig. 10. A diagram of the microcarrier reactor system is shown. It differs from that shown in Fig. 3 by the addition of a settling bottle in the recycle circuit between growth and satellite filter vessels. This sequestering of the cells and microcarriers outside of the agitated volume of the growth vessel enhances formation of cell-bead aggregates and allows essentially cell-carrier free medium to be pumped to the filter. Reprinted from Tolbert and Feder (3).

by Sanford (9). After the cell density had reached approximately 10^6 cells/ml and many of the beads were partially to totally covered with cells, perfusion was initiated. The major difference between this system and the perfusion chemostat which we described earlier was the addition of a settling bottle in the recycled line between the growth reactor and the satellite filter vessel. During operation of the recycled pump, suspension with cells and microcarriers was continuously removed from the culture reactor to the settling chamber. Here, a relatively dense slurry of cells and microcarriers was allowed to settle and aggregate during this temporary residence outside of the vessel's agitated volume. Essentially cell and microcarrier-free medium continued in the recycled circuit to the satellite filter vessel where portions were removed to the effluent reservoir as programmed and the remainder returned to the culture vessel. Fresh medium was added to the growth vessel to replace expended medium removed through the filter. During the settling procedure, beads and cells at high density were brought into close contact. This close contact between the essentially confluent microcarriers stimulated bridging of cells from one carrier to another and eventually formation of large aggregates of microcarriers, as shown in Figure 11. The perfusion of the medium through the settling bottle was important during the formation of the aggregates to prevent any local limiting environment of medium and metabolic products. As the density of cells increased, the perfusion became more and more important to maintain an optimum cell environment. The bridging of the cells between beads resulted in formation of large aggregates which increased the cell numbers significantly beyond that projected for the nominal bead surface area available. There was another unexpected advantage obtained by the highly aggregated condition. The replacement of cell-cell contact by cell-bead surface contact resulted in an easy release of the cells from the beads by short enzymatic treatment in a highly viable condition. After maximum growth of the cells, reaching a cell concentration of approximately 10^7 cells/ml, the medium was removed and the bead aggregates washed three times with PBS containing 0.02% EDTA and then incubated at 25 rpm agitation for 10 minutes at 37° C, with 0.5 grams per liter of porcine trypsin, 0.2 grams of EDTA in Hank's balanced salt solution. When this was done at a growth stage before the formation of the large cell-bead aggregates, the cells were very poorly removed from the beads as has been reported by other workers. However, after the formation of cell-bead aggregates, this treatment procedure completely removed all of the cells from the beads. Consequently, it was

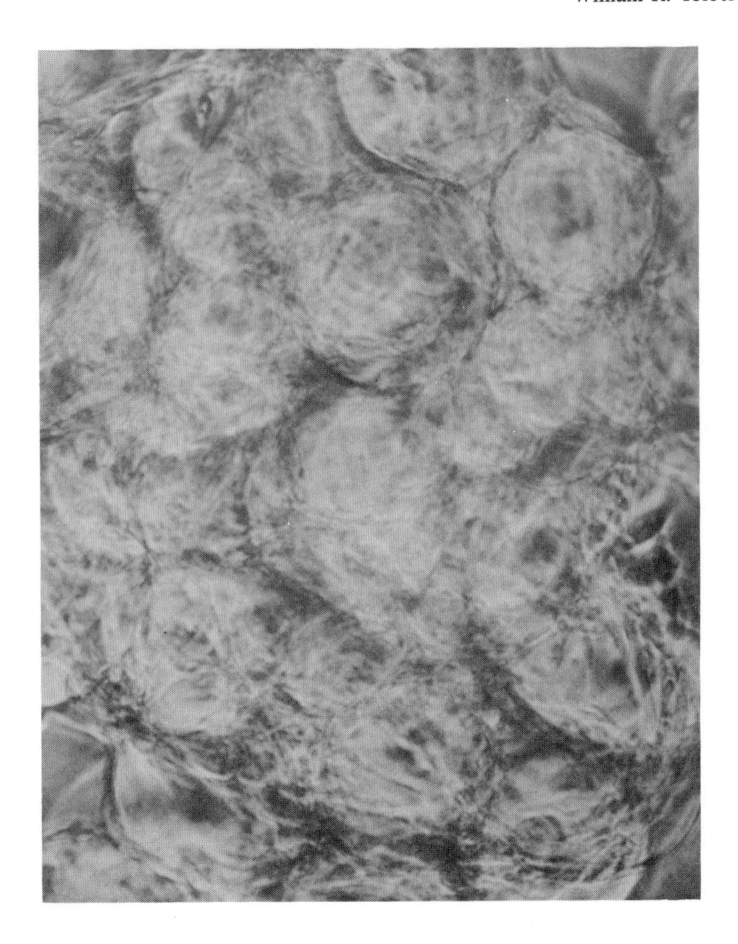

Fig. 11. Photomicrograph of human diploid foreskin fibroblasts are shown on Bio-Carriers after 15 days growth, 80X. Reprinted from Tolbert and Feder (3).

possible to produce an inoculum in a 4-liter microcarrier reactor of about 4×10^{10} human diploid foreskin cells (equivalent to 1,300 690 cm^2 roller bottles) and then to remove the cells and inoculate a 44-liter reactor containing 400 grams of biocarriers with an actual surface area of 188 m^2. A total of 3.4×10^{11} cells was produced in this reactor.

The growth curves and various metabolic parameters for the growth of the human foreskin fibroblast cell, AG1523, in both the 4 liter and 44 liter systems is shown in Figure 12.

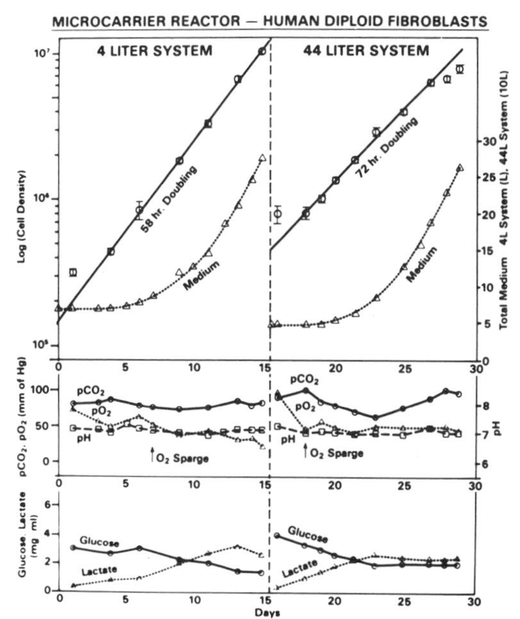

Fig. 12. Growth curves and additional data are shown for 4 and 44-liter microcarrier reactor runs. The 4-liter reactor was used to inoculate the 44-liter reactor at day 15.4. See text for further description. Reprinted from Tolbert and Feder (3).

Table II compares the production of the human foreskin cells in the roller bottle with both the 4 liter reactor. In addition to the valuable, large-scale equivalent production in these reactors, the cell yield gave an efficiency of almost fourfold for the perfusion microcarrier reactor over a conventional roller bottle type of growth system. The human foreskin cells grown in this system were used to produce normal angiogenesis factor and a plasminogen activator. Aliquots of the cell-bead aggregates were also shown to be useful in the production of human fibroblast interferon by the superinduction method. The system seems to solve many of the problems previously reported for use of microcarriers at large scale. It allowed efficient scaleup from smaller to larger microcarrier cultures, even for the sensitive human diploid fibroblast; it provided for cell densities of five to tenfold higher than that reported earlier with significantly higher efficiency in the utilization of expensive medium containing serum.

Recently a uniquely designed perfusion reactor system has been developed to provide maintenance of viable animal cells over long time periods at high cell densities and to provide a method for continual harvest of secreted biomolecules. This

Table II. Microcarrier Reactor: Production of Normal Human Cells

	Surface area (m^2)	Total cells produced	Equivalent roller bottles	Medium efficiency
1 Roller Bottle	0.069	$3x10^7$	1	1
4-Liter Reactor (Average of 5 Runs)[a]	21.6	$4.0x10^{10}$	1,300	3.9
44-Liter Reactor	188	$3.4x10^{11}$	11,000	4.3

[a] Two runs made with 5% bovine calf serum in place of 10% fetal bovine serum used elsewhere.

system provides a means of immobilizing either anchorage-dependent or independent cells at densities from 10^7 to 10^8 cells/ml, allowing continual perfusion with fresh medium to replenish nutrient components and remove metabolic wastes and secreted biomolecules. An efficient system for maintaining dissolved O_2 and CO_2 concentrations is also provided. Figure 13 shows a diagram of this static maintenance reactor (SMR) which utilizes porols tubes to efficiently perfuse the entire volume of the vessel containing cells at high density imbedded in a non-toxic matrix material. During operation fresh medium is continuously perfused through the reactor and cells maintained for one to four months at a time producing desired biomolecular products. These reactors are available in a size range from 10 ml through 16.5 liters. The larger version has a capacity of one to two kilograms of mammalian cells (equivalent to 1,000-2,000 liters of suspension culture).

Figure 14 shows the maintenance of the 9.2.27 hybridoma which produces monoclonal antibody to an antigen isolated from the human melanoma cell line M14. The glucose and lactate levels were maintained at a relatively constant steady-state level over a period of nearly two months. A linear perfusion rate of about 16 liters per day for $2x10^{11}$ cells (80 mls/10^9 cells x day) was maintained over the entire period. The initial 185 liters of Dulbecco's modified MEM (DMEM) medium contained 1% fetal bovine serum which was reduced 0.5% for the next 400 liters and then further reduced to 0.25%. Monoclonal antibody

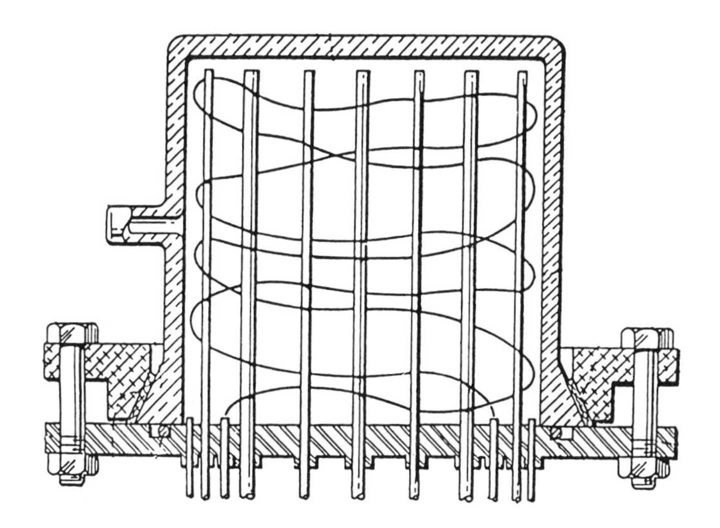

Fig. 13. Diagram of cross-section of static maintenance reactor (SMR) used to maintain cells in non-proliferating state approaching 1/10 tissue density for harvest of secreted products.

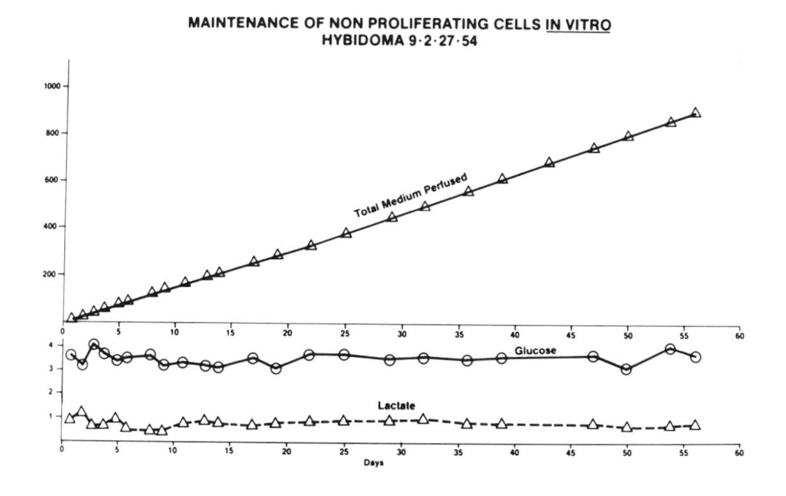

Fig. 14. Maintenance of hybridoma line producing monoclonal antibody for two months with serum concentration as low as 0.25% without antibiotics.

Fig. 15. Production scale 16.5 liter static maintenance reactor automated with computer control of culture parameters.

production was at an average level of 30 mg/liter in both the 0.5% and 0.25% fetal bovine serum-containing medium. Thus the monoclonal antibody production could be maintained by a high density population of hybridomas with perfusion of medium even at very low concentrations of serum. Recently a 3½ month run of the 16.5 liter production scale SMR was completed producing thousands of liters of cell conditioned medium (Figure 15).

In conclusion, perfusion systems provide a means by which to establish a steady-state environment for optimal growth of mammalian cells in culture. Conventional non-perfused systems allow cell densities near 1/1000 of tissue density. The agitated perfusion chemostat and microcarrier systems allow 1/100 of tissue density and the static maintenance reactor allows perfusion of essentially non-proliferating cells approaching 1/10 of normal tissue density. This latter system is particularly useful for long-term secretion of products from more differentiated cells.

REFERENCES

1. Kruse, P., Jr., in Growth Nutrition and Metabolism of Cells in Culture, G.N. Rothblat and V.J. Cristofalo, eds., Academic Press, New York, 1972, p. 14.
2. Himmelfarb, P., Thayer, P.S., and Martin, H.E., Science 164:555 (1969).
3. Tolbert, W.R. and Feder, J., in Annual Reports on Fermentation Processes Vol. 6, (G.T. Tsao and M.C. Flickinger, eds.), Academic Press, New York, 1975, p. 35.
4. Tolbert, W.R., Feder, J., and Kimes, R.C., In Vitro 17:885 (1981).
5. Tolbert, W.R., Hitt, M.M., and Feder, J., U.S. Patent No. 4,289,854 (1981).
6. Tolbert, W. R., and Feder, J. Biotech. Bioeng. 24:1885 (1982).
7. McLimans, W.F., Blasmenson, L.E., and Repasky, E., Cell Biol. Int. Rep. 5:653 (1981).
8. van Wezel, A.L. Nature 216:64 (1967).
9. Sanford, K.K., Earle, W.R., Evans, V.S., Waltz, J.K. and Shannon, J.E., J. Natl. Cancer Inst. 11:773 (1951).
10. Tolbert, W.R., Schoenfeld, R.A., Lewis, C., and Feder, J., Biotech. Bioeng. 24:1671 (1982).

DISCUSSION OF THE PAPER

DR. W. S. HU (University of Minnesota, Minneapolis, MN): You have shown us nicely that you have a few systems to grow cells which all have a common future, that is perfusion. Can you tell what criteria you used to set the flow rate for those systems and do you have a feedback loop to control the flow rate or not?

DR. W. R. TOLBERT (Monsanto Company, St. Louis, MO): We don't have a feedback loop. The method to decide what flow rate to use is trial and error, essentially. We found that in the perfusion chemostat systems, it is proportional to some extent to the rate at which the cells grow. For example, our rodent/rat cells with a 17 hour doubling time required about 10 ml of medium/ml of packed cells per hour, whereas the liver cells that I showed you have about a 50 hour doubling time and require only about 5 ml of medium per hour/ml of cells in the reactor.

DR. J. FEDER: Theoretically you would like to maintain a steady state saturating concentration of all components that are necessary for growth and maintenance. This, of course, requires detailed knowledge of Km's for glucose, amino acids, growth factors and the like. The stead state concentrations to be maintained would be determined from this information. This steady state concentration might be significantly lower than that required in a non-perfused static system. As an example, we generally observe maximum growth rates in perfused cultures at much lower serum concentrations than that required in non-perfused culture. There also are metabolic products which are toxic for the cell, as mentioned by Dr. Tolbert. Ammonia might be such a candidate, but most are not identified. One would like to establish steady state concentrations which are below the concentrations which are toxic for the cells. The system described allows one to establish some homeostatic environment for optimizing all of these variables for growth or cell maintenance and product expression.

DR. W. S. HU (University of Minnesota, Minneapolis, MN): I can understand that when you find out a flow rate, whatever method, you can apply the flow rate to your system. It's easy when you have a steady cell concentration...you just have constant flow rate. But in some of your experiments you essentially have a cell growth, too, in your reactor. So your flow rate must proportionately increase by some number. Then how do you control it?

DR. W. R. TOLBERT: In the perfusion chemostat system, that's what I mentioned. One has to maintain the perfusion flow proportional to the number of mls of packed cells in the vessel. So as the cells grow, you have to increase the flow rate. I also mentioned that that becomes unreasonable after awhile, in which case we start the chemostat mode so you hold the cells also constant. As I said, the ways that we determine how to do that has been empirical. Since these cells actually grow very slowly, it's only necessary to make a change once or twice a day in the flow rate in order to maintain this kind of control. So we don't have a continuous change of the flow rate if that's what you're asking.

DR. NODELL (DuPont): How do you go about selecting the type of microcarrier to use for a given cell type and how do you encourage cells to form bridges between microcarriers?

DR. W. R. TOLBERT: Again, the way we did this is to try as many of them as we could get our hands on and find one that we liked. The one that we liked when we did most of this work is not the same one we're using now, because there has been a major advance in the types of microcarriers available. The types of microcarriers that are now available include those that are solid collagen beads so that just a slight trypsin will dissolve the bead and you have the cell separated from it. In addition, the cells seem to grow much better on this collagen surface. There are also beads that have a collagen coat for the same reason. These were not available at the time we were doing our original work. The way we enhance the bridging is in that settling bottle arrangement on the reactor. We're bringing the cells on the beads into close contact with each other, removing them from the region of agitation of the vessel and allowing them to slowly fall back into the reactor. What happens is the cells completely coat the beads, then it's the cells actually attaching to each other that forms the bridges and it forms mostly in the region where we're letting these beads settle out. It's better to do it that way than just let the beads settle all the way to the bottom of the reactor. Because as you reach higher densities of cells, letting them settle to the bottom for even short periods of time will put a great deal of stress on them, using up all the nutrients and all the oxygen that's available. So in this system we allowed medium to flow by the cells during the entire time that this bridging is occurring.

DR. R. C. DEAN, JR. (Verax Corporation, Hanover, NH): I don't understand the difference between your new cell, Bill, and the Amicon type cell we heard described this morning by Bio-Response. Can you tell us what is different?

DR. W. R.TOLBERT: Well, there are several differences. The major difference is the cells are not growing on these porous tubes. Now these tubes are really porous, they have pore size in the range of microns and tens of microns. The cells are being either attached to microcarriers and not growing or contained in the interstices between the matrix material with which we fill the reactor. It is almost like a chromatography column filled solid with mass. There is no agitation at all and of course there is no sheer, or very, very little sheer, due to the movement of medium throughout the reactor.

DR. P. BROWN (Bio-Response, Inc., Hayward, CA): I didn't understand in that last system what you did with the continual growth of the cells. Are they now growing?

DR. W. R. TOLBERT: They are not growing. They are being maintained to a large extent.

DR. P. BROWN (Bio-Response, Inc., Hayward, CA): What's maintaining them, the physical constrictions or what?

DR. W. R. TOLBERT: In the case of anchorage dependent cells, one of the major ways of stopping these cells from growing is to let them run out of surface area on which to grow and also by reducing nutrient levels. In the case of non-attached cells, reducing nutrient levels or other parameters are used to control the growth. In fact, there is some small growth occurring throughout the reactor. But the point I want to make is that this is not a growth system. We put in the number of cells that we want to be there in a production mode, so if we want a kilogram and a half of cells, we have to produce that amount in another system, a growth system, and the inoculate them into our maintenance system.

DR. P. BROWN (Bio-Response, Inc., Hayward, CA): How's the viability with time in that kind of system?

DR. W. R. TOLBERT: We can't sample very well because of the way it's designed. However, we look at other kinds of parameters like oxygen uptake, CO_2 evolution, product production, and these things seem constant over long periods of time.

DR. P. BROWN (Bio-Response, Inc., Hayward, CA): Have you looked at LDH levels?

DR. W. R. TOLBERT: I haven't done that directly, but I'm not sure if other people have yet or not.

DR. B. K. LYDERSEN (Research and Development Laboratory, K C Biological, Lenexa, KA): I wonder with this matrix, first off, I guess you really can't tell us something about what it's made of.

DR. W. R. TOLBERT: Well, I told you in the case of the anchorage dependent cells it was microcarriers.

DR. B. K. LYDERSEN (Research and Development Laboratory, K C Biological, Lenexa, KA): Oh, just microcarriers?

DR. W. R. TOLBERT: Well, one might add other things. But essentially, the matrix material is finally divided non-toxic material.

DR. B. K. LYDERSEN (Research and Development Laboratory, K C Biological, Lenexa, KA): I was just curious about this idea that the cells are not dividing. I think it's a good idea but is it possible that you really do have the same amount of growth that you have in a flask or some other system and those cells die and others take their place or can you really verify that?

DR. W. R. TOLBERT: I can't verify it completely, but to some extent that probably is occurring. I think it's occurring at a much lower rate as it does in our bodies. There is some turnover of cells in our bodies, but it's a low rate. We really have just scratched the surface of what may be done eventually in a system like this, where one can look at differentiation agents and so forth.

DR. B. K. LYDERSEN (Research and Development Laboratory, K C Biological, Lenexa, KA): So, I guess the other point is, could you use the same medium condition, same condition somehow, say in a flask or a spinner with a filter and get the same kind of results or is this something unique that's going on?

DR. W. R. TOLBERT: I don't believe any system which allows even the reduced amount of agitation as with some of our gentle sails reactors would work. The agitation and sheer in those, I think, would prevent this long-term kind of maintenance of the cells. I think you can see this effect by just putting cells in T-flasks versus trying to maintain cells a long time in a spinner. If the cells' growth drops in the spinner, the cells all die out. If a cells' growth drops in a T-flask, they still usually maintain for quite awhile.

CELLULAR AND BIOCHEMICAL ASPECTS OF HUMAN TUMOR CELL GROWTH AND FUNCTION IN HOLLOW FIBER CULTURE

M. C. Wiemann
B. Creswick
P. Calabresi

Roger Williams
General Hospital
Providence, RI

J. Hopkinson
R. S. Tutunjian

Amicon Corporation
Danvers, MA

A prototype process-scale hollow fiber cell culture system was evaluated for its ability to support the growth and maintain the function of large numbers of cells for long periods of time. Culture cartridges contained 1,000 hollow fibers with a luminal surface area of 1,000 cm^2 and a cutoff of 10K or 50K daltons, and had an extra capillary volume of 25 cm^3. About 5 x 10^7 DLD-1 human colon carcinoma cells were inoculated into the cartridge. One liter of culture medium containing 10% fetal bovine serum was recirculated at a rate of 100 ml/min and full volumes were replaced every other day. Glucose consumption, used as an index of cell growth, reached a plateau of 1 gram/day during the third week of culture. Culture fluid, harvested from the cartridge every four days, contained increasing levels of the enzymes LDH and GOT and the glycoprotein CEA, and negligible amounts of serum proteins. Culture cartridges terminated at six weeks held 2.5 x 10^9 cells with a karyotype similar to the inoculum. Histopathology revealed that cells had infiltrated the fiber wall matrix and filled the interfiber spaces, growing there as a dense, moderately-differentiated colon carcinoma which resembled the original DLD-1 human tumor.

DISCUSSION OF THE PAPER

DR. R. C. DEAN, JR. (Verax Corporation, Hanover, NH): You said you were, in the last slide, comparing apples to apples, but you didn't tell us the concentration of the antibody in the cell suspension liquor. You've heard today concentrations all the way from 10 micrograms/ml up to 1000 micrograms/ml from suspension cultures.

DR. J. HOPKINSON: The concentration of antibodies in that case was around 10 micrograms/ml.

DR. R. C. DEAN, JR. (Verax Corporation, Hanover, NH): That's very, very low for a free-cell suspension.

DR. J. HOPKINSON: It's very, very concentrated, too, when it gets into some of the other systems. Some hybridomas that may perform very well in ascites fluid will not do well in a suspension system and vice versa.

CONTINUOUS PRODUCTION OF MONOCLONAL ANTIBODIES BY CHEMOSTATIC AND IMMOBILIZED HYBRIDOMA CULTURE

Subhash B. Karkare
Philip G. Phillips
Deborah H. Burke
Robert C. Dean, Jr.

Verax Corporation
Hanover, NH

This paper describes progress at Verax toward monoclonal antibody (MAb) production by continuous tissue culture. Hybridoma cells were grown at steady state in chemostat culture at various dilution rates in order to determine cell concentration and MAb productivity. Cell concentrations as high as 4×10^6 cells/ml were obtained with MAb space-time productivity of about 8 mg/ℓ-hr. Antibody titres greater than 300 μg/ml were reached. The cultures were further optimized with respect to pH, temperature and dissolved oxygen in order to maximize MAb productivity. The same hybridoma cells were immobilized in a proprietary carrier matrix. Cell densities as high as 7×10^7 cells/ml of matrix were observed. Proprietary medium improvements lead to a significant increase in cell productivity. Compared to conventional batch tissue culture, the space-time productivity of MAb with immobilized-cell culture in commercial production is expected to be at least tenfold greater. A review and comparison of immobilized-cell culture methods is offered. The virtues of fluidized slurries of fibrous bead carriers are elucidated.

I. OBJECTIVES

The objective of this paper is to present a progress report on Verax's development of high-performance tissue culture methods for producing monoclonal antibodies. A second objective is to review the state-of-the-art of monoclonal antibody (MAb) production, including the performance of the leading methods.

A. Relevance

Table I is a projection of market demand for MAb to 1990. We have surveyed numerous projections and have concluded that Weinert's (1983)[1] is as valid as can be -- when projecting markets that do not exist for a product that does not exist in significant quantity. Note that Table I is not specific about animal/animal, animal/human or human/human hybridomas for making animal or human MAbs. However, one can be quite certain that all of the human-therapy applications and perhaps many of the immuno-affinity separation (1ASep) applications will demand human MAbs because of the allergic reactions of some people to extra-species proteins. Today MAb from mouse/mouse hybridomas is selling for $2,500-10,000 in >1 gram lots for applications other than in vitro diagnostics. For that application, antibodies cost up to $millions/gram. However, the in vitro diagnostic market uses such small quantities (e.g. one mouse may load some 10,000 kits) that it is of no interest to the MAb mass producers. For the other applications in Table I, human antibodies are desired, but generally are not available now. When they become available in the next few years, it is anybody's guess what their initial cost will be. We expect that all human MAbs will be produced in vitro, even though it may become possible to generate them in animals This is because the cost of separating the animal's own antibody from the human MAbs probably will be prohibitive, and the consistency of the output probably will be insufficient. Below we shall consider only in vitro mass production of >1 g lots of MAbs. Lots as large as several kilograms are envisioned for 1990.

---

[1] see References

Table I. U.S. market for monoclonal antibodies ($ millions/year)

Year	In Vitro Diagnostics[1]	Radio Imaging	Immuno-Chemotherapy[2]	Passive Immunity	Thera-peutic	Separation Purification	Total
1983	59	0	0	0	0	--	59
1985	150	25	50	15	50	22	312
1990	500	250	400	150	500	100	1,900

[1]kits and reagents.
[2]and control of immune response.

SOURCE: Boston Biomedical Consultants, Inc.

1. Conventional *In Vitro* Production

From a typical batch/free-cell culture, one can expect on the average about 0.5 g/ℓ gross output in 10 days or about 2 mg/ℓR[1]/hr average space-time productivity. From a chemostat/free-cell culture (based on Verax data) we expect 6 mg/ℓR/hr, or about three times the batch/free-cell culture average. This higher performance results because the steady-state chemostat's operating condition can be optimized precisely in order to enhance MAb production. For example, optimum nutrient concentration is easy to maintain in the chemostat, but impossible to maintain in the batch culture.

On the basis of these figures, one can generate Table II which shows the size of the conventional production capacity needed to satisfy the U.S. MAb market until 1990. In preparing Table II, we have assumed that the 1984 average price of $3,750/gram will drop linearly to $500/gram in 1990. We note that somewhere between 60,000 and 180,000 liters of reactor volume will be needed in 1990 to meet the U.S. market demand (and 2-3 times that production capacity for the worldwide market). At an average cost of $2,000/liter, this means the world will have invested over $1 billion in MAb production capacity. For those of us who are expecting to be large

Table II. MAb production capacity projections batch and continuous chemostat free-cell-suspension culture

				Production Capacity[1] Required (liters)	
Year	MAb Price ($/gm)	U.S. Market Volume ($millions)	U.S. Production[2] Output (kg/yr)	Batch	Continuous Chemostat
1985	3200	312	98	6.2×10^3	2.1×10^3
1990	500	1400	2800	1.8×10^5	0.6×10^4

[1]excluding *in vitro* diagnostics (see Table I).
[2]330 days/year x 24 hours/day.

[1] ℓR = liters of reactor active volume

factors in the market, this capital cost is alarming. Plainly MAb producers are impelled strongly to find means to improve grossly the productivity of their invested capital (and labor too).

2. Review of Various MAb Production Processes

Table III-A and III-B present performance estimates for various old and new processes for mass producing antibodies.

Mouse MAb Production on the basis of our analysis (which appears to be consistent with the literature and private reports) requires a 100 mouse colony to be maintained in order to produce 1 gram/month of MAb or 15 μg/hr/mouse in the colony. Of this colony, some 30% of the mice are in a preparatory phase before injection with hybridoma. After about six weeks, the mice die from the trauma of their peritoneal ascites. For a simple comparison, we used in Table III-A the definition:

one "Mousepower" = 15 μg of MAb/hour

Available cost data for manufacturing MAbs from mice is very unreliable. Our economic analysis and private information from some manufacturers indicates purified MAb costs between $5,000 and $10,000/gram when all costs, including overhead and downtime, are considered. There are few economies of scale in dealing with mice, although some of the keepers of very large colonies (they now exceed 30,000 mice) are automating feeding, cleaning, etc. However, we do not expect significant cost reduction to be possible with mouse colonies.

Batch/Free-Cell Suspension Culture has not been improved significantly in recent years. Obviously, improvements in medium, use of genetic enhancers, and discovery of computer algorithms for better control could increase productivity and yield. We are critical of batch/free-cell suspension culture because it is inherently a transient process, which can never be optimum. Further, it requires growth of a new cell colony for every batch and the eventual wasting of most of the hybridoma biomass produced. While many workers are attempting to develop computerized-control algorithms for batch/free-cell suspension tissue cultures, in our opinion these attempts are wasted on an inefficient process. Typical batch-culture cell concentrations are the order of 10^6 viable cells/ml with productivity of 1 μg/10^6 viable cells/hr and 80% cell viability.

The Chemostat/Free-Cell Suspension method has advantages over batch/free-cell suspension culture, as demonstrated in Table III-A. According to our experience, it is a tissue culture much easier to operate than a batch because of its continuous and steady-state nature. However, it also is inherently a growth situation because hybridoma biomass is wasted

TABLE III-A. Comparison of MAb production processes

System	Productivity [g] Mousepower/ℓ	Productivity [g] mg/ℓR[a]/hr	Yield [b] g/ℓM	Harvest Concentration [e] g/ℓH	Reference
A) Mouse (15 μg/hr)	6	9×10^{-2}	na	10 [f,c]	Fanger (1983)
B) Batch/free-cell	133	2	-	-	Phillips (1983)
C) Chemostat	400	6	0.3	0.3	Verax Data, VX-2
D) Chemostat with cell return	400	6	-	-	Lewis *et al* (1984)
E) Estimated Microtubule[h]	3,000	45	-	-	Inloes *et al* (1983)
F) Encapsulated HY[d]	174	2.6	0.5	2.5	Littlefield *et al* (1984)
Projected CF-IMMO					
pessimistic	730	11	0.5	0.5	Verax Data, VX-2
optimistic	15,000	226	1.0	1.0	Verax Data, VX-2

[a] ℓR = reactor volume (liters)
[b] ℓM = medium volume consumed (liters)
[c] ascites
[d] microshell encapsulated HY = hybridoma
[e] ℓH = harvest liquor volume (liters)
[f] typical; ranges from 1 to 20
[g] gross output; not accounting for purification loss (~30%).
[h] calculated for hybridomas densely packed at 1.3×10^8 viable cells/ml and producing 0.5 μg/10^6 viable cells/hr.

TABLE III-B. Monoclonal antibodies. Comparison of Verax process to other manufacturing processes (all values relative to Verax values)

		OTHER PROCESSES					
FEATURE	VERAX	A	B	C	D	E	F
Productivity	1	0.0004	0.01	0.03	0.03	0.20	0.01
Reactor Size	1	2500	100	30	30	5	100
Capital Cost	1	12	10	30	30	8	50
Purify Cost	1	12	6	4	4	2	3
Labor Cost	1	7	4	3	1	1	4
Nutrient Cost	1	1	2	2	1	1	1
Total Cost of Product	1	20	15	10	10	4	15

MANUFACTURER	PROCESS USED
Celltech (UK)	B,C
Monsanto/Invitron	D,E
BioResponse	E
Damon Biotech	F
Karyon	F
Hybritech	A,B
Endotronics	B
KC Biologicals/Corning	D
Charles River Breeding	A

continuously in the harvest stream. This biomass production consumes expensive medium and may be unnecessary. We have achieved cell concentrations up to 4.5 x 10^6 cells/ml, although stable concentrations around 2 x 10^6 for long periods are more typical. Productivities of 1 μg/10^6 viable cells/hr are typical of the several hybridoma/MAb combinations we have tested.

Free-Cell Chemostat with Filtered Cell Return is a technique pioneered by several groups, namely Himmelfarb et al (1969) of Arthur D. Little, Inc. and Feder and Tolbert (1983) of Monsanto. A spinning filter in the center of the reactor serves essentially as a tangentially-washed filter and prevents escape of the cells. Stationary tangentially-washed filters have also been used for this purpose. The harvest liquor, containing the antibodies, unspent medium components, etc., is drained through the filter; the separated cells are washed off into the reactor. By this approach, hybridoma cell concentrations between 3.8 x 10^6 cells/ml (Lewis et al 1981) and 2.5 x 10^7 cells/ml (Feder and Tolbert 1983) have been attained. Productivities reported are of the order of 6 mg/ℓR/hr of MAb. Whether these are typical or peak values is unstated.

Microtubule Bundle in a Shell, the shell-and-tube bioreactor, has been available commercially for years (e.g. the Vitacell, Amicon Corporation). Hybridomas are grown into a solid mass on the shell side (Calabresi et al 1981) and nurtured by medium components diffusing through the porous tubules. The MAbs are flushed from the shell side or, if the tubule pores are large enough to pass 150 KD to 1,000 KD molecules, the MAb's may pass into the tubules' lumens.

Bioresponse Corporation (1983) reports commercial production using lymph fluid from an attached living cow as nutrient. Scant data is available to us on performance of such a bioreactor. We have used the cell packing densities of Inloes et al (1983), scaled from 5 μm *S. cerevisiae* yeast (3.5 x 10^9/ml) to 14 μm hybridomas (1.6 x 10^8/ml), and assumed productivity of 0.5 μg/10^6 viable cells/hr, 80% viability and a shell-side volume of 70% of total reactor volume to get the performance estimate in Table III-A. The estimated bioreactor space-time productivity exceeds all other methods by a large factor. This estimate may be high because of diffusion resistance to medium transport among the cells, but we expect that it is not far out of line.

The defects we find with the shell and tube approach are the very high cost of these reactors, even scaled-up to commercial-production size, and the fact that cells cannot escape to relieve the pressure of colony growth. Inloes et al (1983) report that the growing yeast colony broke through certain types of tubules in a few days. Such short life of the reactor would be discouraging, but might be improved by choice of proper tube materials. In the end, making a cell immobilizing and mass transfer surface from porous tubules (at approx. 4 m^2/ℓ) must be much more expensive than making the large surface area (approx. 40 m^2/ℓ) of a thick slurry of 200 μm carrier beads as in the Damon Encapsul and the Verax CF-IMMO processes. So despite the shell-and-tube bioreactors high performance in Table III, we are of the opinion that it is not a strong competitor for a commercially-successful production process.

Microshell Immobilization: this technique has been pioneered by Lim, et al (1981) and commercialized by Damon Biotech Inc., who has patented the process of encapsulating hybridomas in very thin polylysine shells of 200 μm diameter. We have seen various performance numbers on this process, e.g. see Littlefield et al (1984). Apparently Damon is now operating with about 25% microshell volume fraction in their stirred reactors, (Littlefield et al 1984), and running for about 10 days after the culture starts until the hybridomas grow to fill the microshells. Because the hybridomas will eventually break the microshells (Lim et al 1981), Damon's is inherently a batch process, although it may be operated for about 10 days as a chemostat or fed batch. We expect that Rupp (1984) will provide more information on this process at this meeting.

Given the data available to us, we calculated the values shown in Table III indicating a reactor volume productivity of about 2.6 mg/ℓR/hr. The harvest concentration is high inside the microshells, with MAb concentrations reported of 2.5 mg/ml (Littlefield et al 1984). But about 5 times the microshell volume of medium is employed, so the yield is 0.5 mg/ml of medium. Note in Table III that the microshell-immobilization process does not exceed the chemostat in reactor space-time productivity. We believe this is because the microshell immobilization process is inherently a batch process in which the hybridoma population must be grown (at doubling times of 15-30 hours) from the order of 10^5 viable cells/ml to the order of 10^8 viable cells/ml. Because it is a batch process, the culture conditions can never be perfectly optimized, as they can be in a steady-state chemostat. Both chemostat and microshell immobilization processes are inherently growth situations.

Immobilization in Gels has been employed for hybridomas as reported by Nilsson et al (1973) who formed agarose beads, containing the cells, of the order of 100 μm diameter and made by emulsification in oil. Nilsson reported good results, but gave no quantitative data. We have experimented extensively with immobilizing hybridomas in 4% agarose beads of about 200 μm diameter. The results were poor in that the hybridomas did not colonize the agarose beads to very high packing densites. Occasionally when colonies did grow, they expanded as a "solid" tumor which physically ruptured the beads. Productivity and viability were poor; the colonies died within two weeks in a chemostat. It should be possible also to immobilize the cells in a thin layer of gel that is washed tangentially by nutrient. Hybridomas are frequently cloned in soft agarose, which suggests that this technique might be made to work. However, it would be a delicate and probably impractical situation because a soft agarose, which will not rupture on colony growth, has consequentially a low physical strength and great fragility.

Fiber Mat Immobilization has been practiced by Feder and Tolbert (1983) where cells were immobilized in a mat and perfused with nutrient continuously. They supplied no data on cell-packing density in the bioreactor volume, productivity, yield and harvest product concentration. From our experience with such systems, we would expect productivity based on mat volume to increase more or less with the cell-packing density so that an improvement of 10-20 times over free-cell suspension culture could be expected. However, the overall bioreactor space-time productivity might not be so impressively high if the mat only occupies a fraction of the bioreactor. Restricted mass transfer to cells deep in the mat would limit productivity and cell viability. Hybridomas are about 14 μm in diameter. Our experience suggests that when the cells are packed more than 5 deep, the inner cells begin to starve. Inloes et al (1983) reported a 100 μm thick dense-packed layer of viable yeast cells.

Fiber Bead Immobilization is an alternative approach employing an immobilizing-matrix, in the form of beads, in a packed bed or stirred slurry. Even though hybridomas are lymphatic cells that generally do not attach to surfaces, we have successfully entrapped them in fibrous polymer beads about 200 μm in diameter, with internal pore dimensions of the order 40 μm. Other investigators have employed porous ceramic and glass beads in slurries or packed beds, e.g. Messing et al (1979) or porous Celite (John Mansville) particles to immobilized microorganisms. We do not know whether these carriers will work for hybridoma cells.

Figure 1 is an electron micrograph of one of the fibrous structures employed and Figure 2 shows an immobilized hybridoma colony after 22 days of incubation. It has grown to close to the theoretical packing density (2.8×10^8/ml for hard, 14μm balls). We have found in general, with proper fiber structures and materials, especially-tailored medium and particular operating protocols, that we have been able to achieve excellent viability of the cells (80-90% average) and productivities significantly higher than the state-of-the-art. This approach has a special advantage over the microshell encapsulation method in that the expanding hybridoma colony can expel excess cells

Figure 1. Unweighted matrix (pores approx. 40 μm)

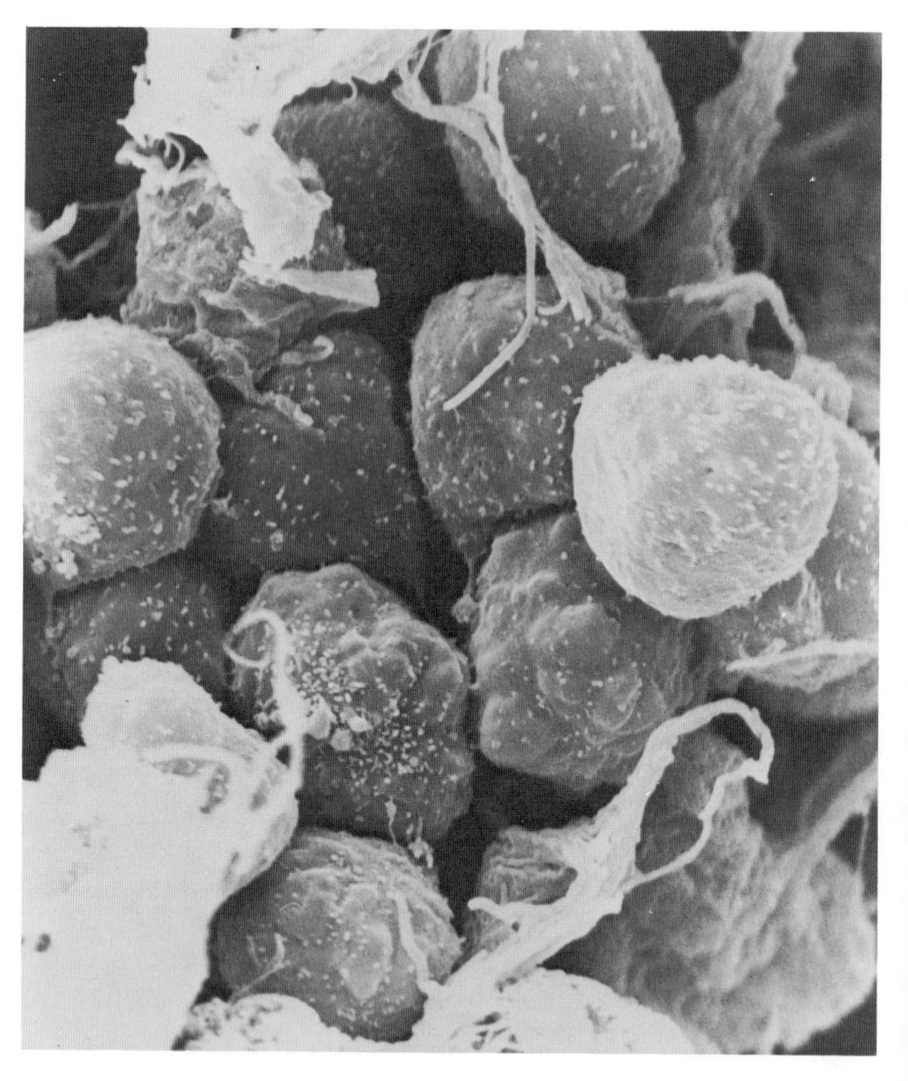

Figure 2. Immobilized hybridomas (cell diameter approx. 14 μm)

through the outer pores of the fibrous beads without rupturing their structure. We have now operated systems using these IMMO beads for several months and have observed no physical damage to the matrix structure. With some structural materials there is biochemical damage, however, we have discovered proprietary materials which avoid this problem. In this fashion, Verax has been able to produce MAb's continuously, in its laboratories, for many months from a single inoculation of hybridomas.

The Verax IMMO beads are used as a thick slurry, with about 40% solids, in a fluidized bed, as will be described below. Another approach is to pack the beads into a fixed-bed column, but our experience is negative because of bead compression, flow channeling and growth of the hybridoma colony to fill the entire reactor, causing low-mass transfer rates and high pressure drop.

As shown in Table III, we expect this Verax CF-IMMO system to demonstrate high space-volume productivity, high yield and acceptable product concentration in the harvest.

3. Verax Continuous Immobilized-MAb (CF-IMMO) Process

A schematic of the Verax antibody manufacturing process is shown in Figure 3. The IMMO matrix beads are manufactured by a proprietary technique from special materials, both of which are being patented. Figure 4 is an SEM of a weighted-matrix IMMO bead without cells, while Figure 2 shows the hybridoma colony inside an IMMO bead. The cell concentration inside the bead runs about 10^8 cells/ml for hybridomas of 14 μm diameter. In the liquor surrounding the beads, the concentration is about 1/10th that value. The average cell concentration in the bioreactor volume is about 40% of that in the beads or approximately 5×10^7 cells/ml.

The thick slurry is suspended and agitated in a fluidized bed as illustrated in a small laboratory reactor in Figure 5 and schematically for a production rig under construction in Figure 6. Note in Figure 6 that the culture liquor is recycled outside the fluidizing bed in order to produce the fluidization action and to condition the culture liquor at each stage independently. Oxygen is transferred to and CO_2 is removed from the bed by a membrane gas exchanger. An external heat exchanger controls the bioreactor temperature. Instrumentation such as pH, DO_2, temperature, turbidity, etc., probes are installed in the external recycle loop for easy access and calibration. The recycle dilution rate can be very high (we have run up to 600 hr^{-1}), while the medium throughflow dilution rate can be very low (we have run down to 0.006 hr^{-1}). Separation of the recycle and throughflow

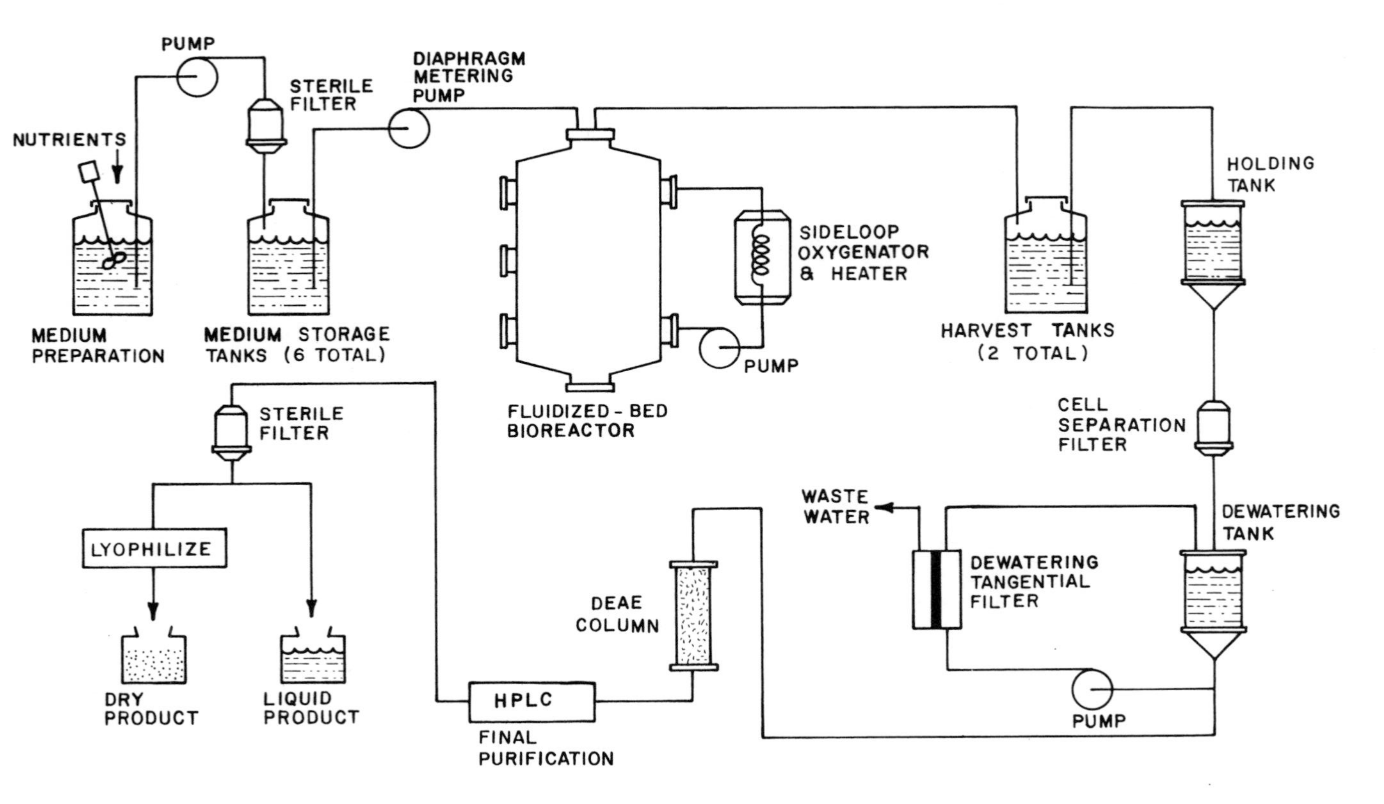

Figure 3. Schematic of MAb Production System.

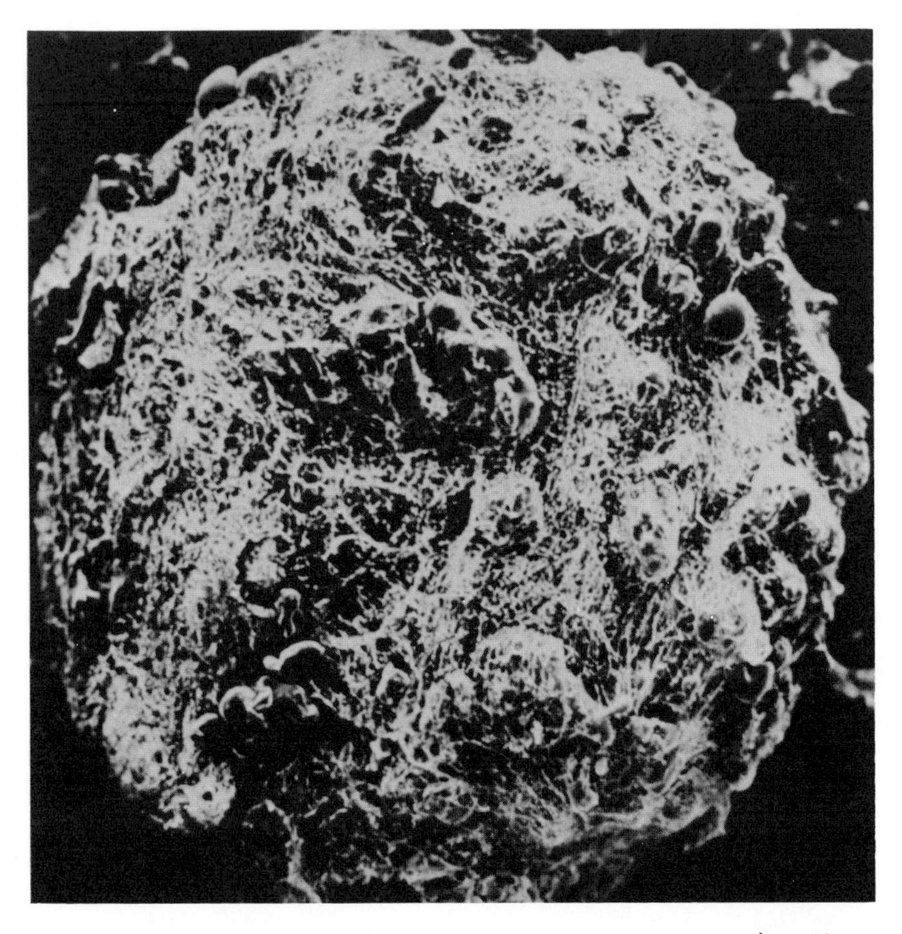

Figure 4. Prototype weighted-matrix carrier bead approx. 400 μm diameter; sp gr ≅ 1.5

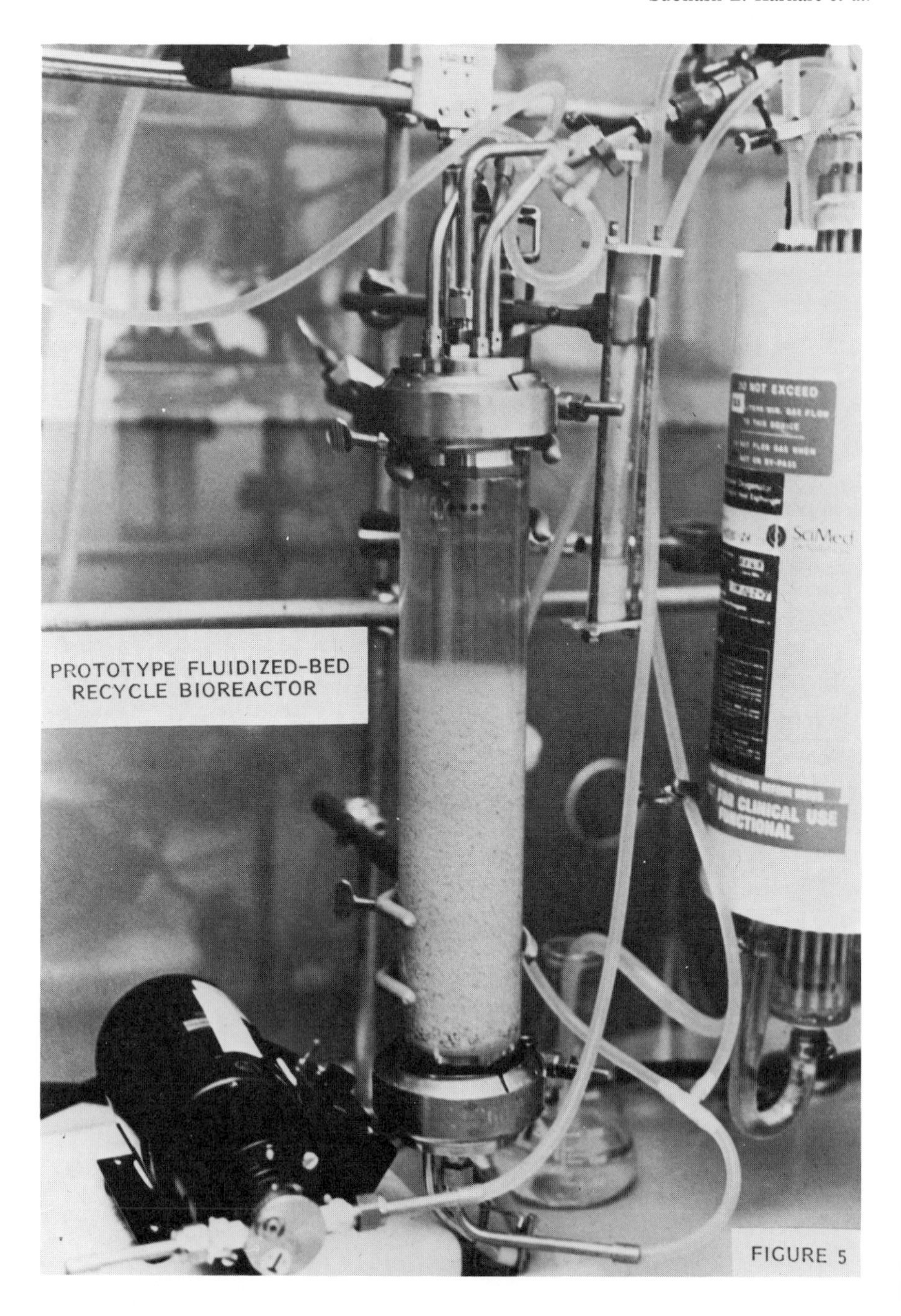

Figure 5. Prototype fluidized-bed recycle bioreactor.

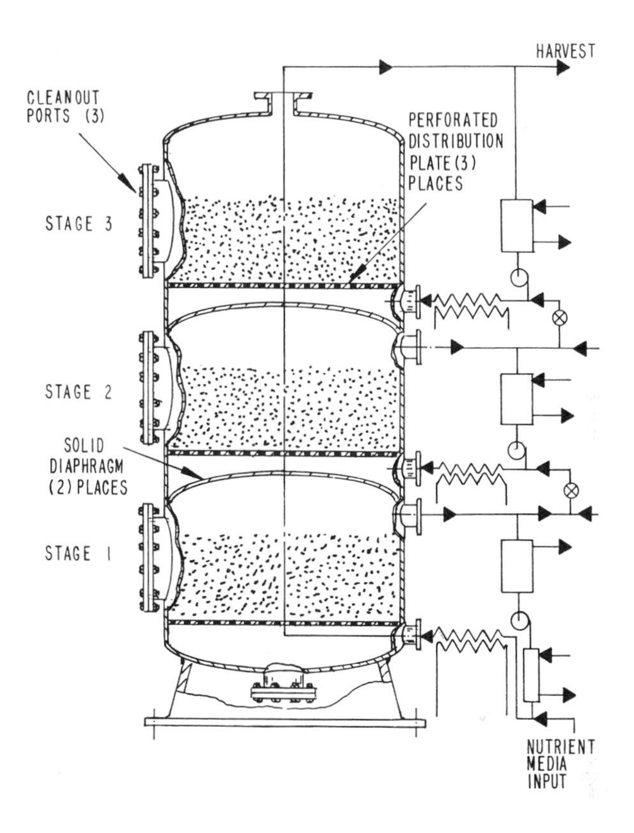

Figure 6. Schematic of cascaded - stage fluidized - bed CF-IMMO Bioreactor

dilution is a very convenient feature of this system. The IMMO matrix beads we employ have specific gravities up to 2.0, so the fluidization velocities (that suspend the bead) are relatively high and of the order of 2 mm/sec.

As shown in Figure 6, individual fluidized and well stirred stages may be cascaded to make a plug-flow-type reactor that is nutrient rich in the early stages and product rich in the final stages.

We have developed dynamic protocols to encourage maximum productivity. Nevertheless, the system runs at approximately steady-state so control is relatively simple. In the Verax Fermentation Kinetics Laboratory, an optimum protocol is developed and proven by extensive testing in mini-fluidized bed reactors, such as shown in Figure 5. The kinetics data (see Figure 7 for a typical sample) are developed so that our engineers have a full appreciation of the operating map in order to optimize the combination of medium, hardware system, operating protocol, and purification procedure. All of these aspects are strongly linked; so an effective, inexpensive and precise means to determine kinetic data is essential to low production cost.

As shown in Figure 3, after the bioreactor, the antibodies are purified by first removing cells and cell debris from the harvest liquor, with a microporous filter, followed by removal of about 95% of the water by tangential-flow ultrafiltration. A DEAE column is next employed to extract bovine serum albumin. That is followed by HPLC final purification before filter sterilization and lyophilization, or the bottling of a sterile-liquid product.

4. Aseptic Operation

Because CF-IMMO cultures may run continuously for many months, it is absolutely essential that perfect aseptic operation be achieved. Verax has operated asepticaly with CF-IMMO for more than four months (>3,000 hours). This achievement is not easy; painstaking attention to detail and especially-engineered hardware and special techniques are necessary. We avoid the use of antibiotics because it is our policy not to inject any substance into the harvest liquor which we must later remove, and in order to prevent protective adaptation of potentially-contaminating organisms. Verax's CF-IMMO systems are _in situ_ steam sterilized with ultra-pure steam (from a Millipore Super-Q feeding a Sybron-Barnstead pure steam generator). Medium is quarantined for at least five days under refrigeration and in the dark while its sterility and nutritional value are validated.

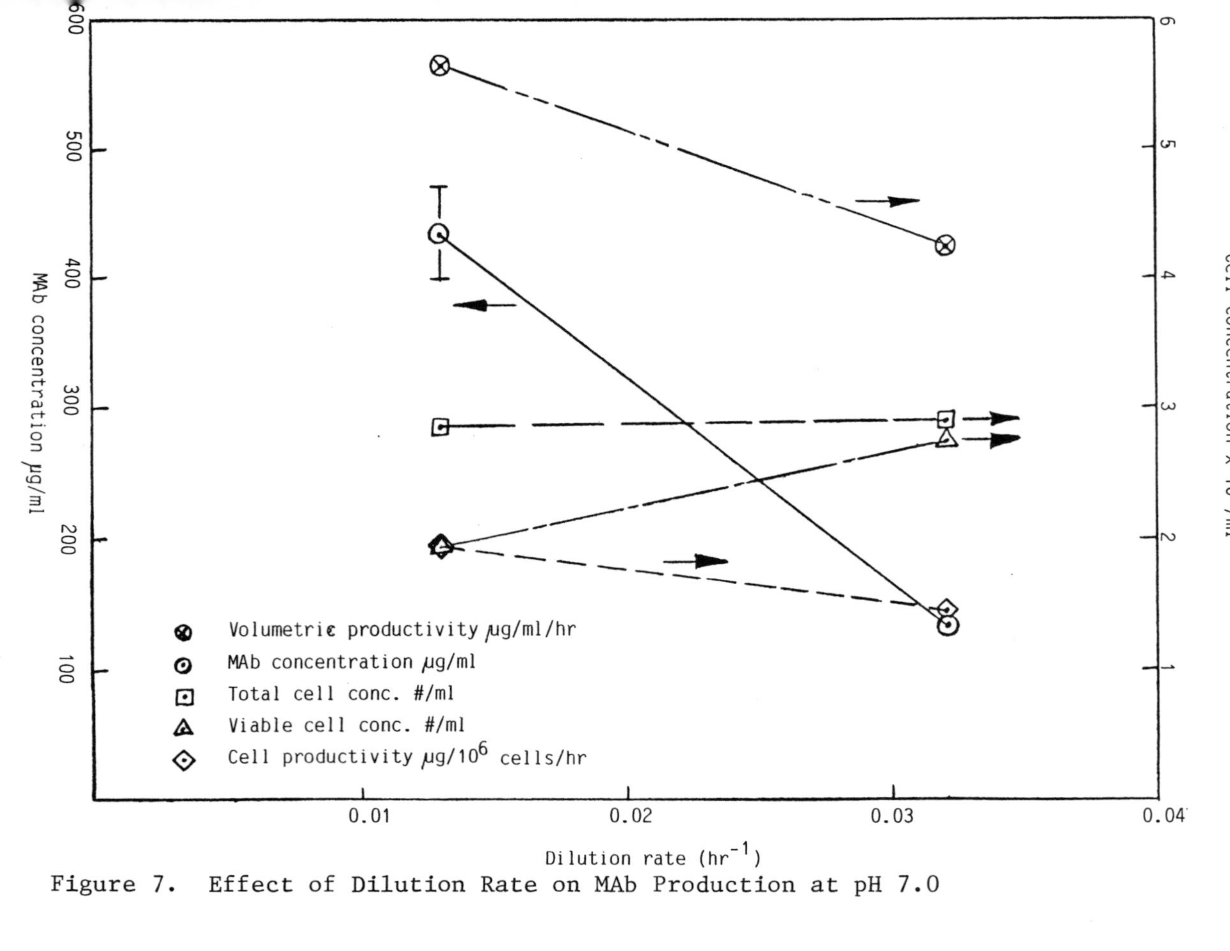

Figure 7. Effect of Dilution Rate on MAb Production at pH 7.0

5. Verax Production Process

The production of MAbs starts with preparation of the hybridoma cells in our Cell Preparation Laboratory. Cells are usually propagated from a number of about 10^3 from frozen stock. Then they are checked for contamination (and decontaminated as necessary), subcloned to select high producers and then cultured up to about 10^9 cells for kinetics measurements in our Fermentation Kinetics Laboratory. This is done in the mini-bioreactors shown in Figure 5. In chemostats, inoculum is grown for our 10-ℓ MAb pilot plant reactors. With a 24-hour doubling time, about three days are required to increase the cell population by 10X. So the process just described requires several weeks time lapse.

Figure 8 is a photograph of the Verax MAb pilot plant. Currently the plant is being validated in the chemostat mode. When it is brought into the CF-IMMO mode in the Fall 1984, it will have a large production capacity for MAbs.

CONCLUSION

In this paper we have attempted to compare several conventional and advanced means for large-scale manufacture of monoclonal antibodies. We have also reported in some detail on our continuous culture with immobilized hybridomas, which appears to give superior results, although it has not yet been proven in sustained commercial production. While some of these advanced systems yield considerable gains over conventional processes, we do not believe they have yet reached to nearly their ultimate, within factors of 10 in cost reduction. We expect that the market price of monoclonal antibodies will drop by orders of magnitude over the next 10-15 years. Instrumental in that drop will be greatly enhanced production processes.

ACKNOWLEDGEMENT

The work reported here was supported by the Verax Corporation with substantial assistance from the Small Business Innovation Research programs of the US National Science Foundation and National Cancer Institute. The research was performed by

Figure 8. Monoclonal Antibody Pilot Production Plant

the authors supported by many numbers of the company's staff, especially Sandra L. Warner, Johanna vG. Sample and Peter W. Runstadler, Jr.

REFERENCES

1. Bio-Response, Inc. (550 Ridgefield Road, Wilton, CT); Annual Report; 1983.
2. Calabresi, P., McCarthy, K.L., Dexter, D.L., Cummings, F.J., and Rotman, B.; "Monoclonal Antibody Production in Artificial Capillary Cultures" (Abstract) Proc. Am. Assoc. Cancer Res., 22:302, 1981.
3. Fanger, M.W. (Dartmouth Medical School); Private Communication; 1983.
4. Feder, J. and Tolbert, W.R.; "The Large-Scale Cultivation of Mammalian Cells"; Scientific American, 248, #1:36-43, 1983.
5. Himmelfarb, P., Thayer, P.S. and Martin, H.E.; "Spin Filter Culture: The Propagation of Cells in Suspension"; Science, 164:555-557, 1969.
6. Inloes, D.S.; "Immobilization of Bacterial and Yeast Cells in Hollow-Fiber Membrane Bioreactors", Ph.D Thesis, Stanford University, 1982.
7. Inloes, D.S., Taylor, D.P., Cohen, S.N., Michaels, A.S. and Robertson, C.R.; "Ethanol Production by Saccharomyces Cerevisiae Immobilized in Hollow-Fiber Membrane Bioreactors"; Applied and Environmental Microbiol., 46, #1:264-78, July 1983.
8. Karkare, S.B., Burke, D.H., Dean, R.C. Jr., Souw, P. and Lemontt, J.; "Kinetic Studies on Production of α-HCG by Living Immobilized Cells of Genetically-Eengineered S. Cerevisiae"; Paper to be presented at the Fourth International Conference on Biochemical Engineering, Galway, Ireland, September 30-October 5, 1984.
9. Lewis, C. Jr., Tolbert, W.R. and Feder, J., "Large Scale Perfusion Culture System Used for Production of Monoclonal Antibodies"; Abstracts from the Third Annual Congress for Hybridoma Research, February 19-22, 1984, San Diego, CA, Hybridoma 3, #1:74, 1984.
10. Lim, F. and Moss, R.D.; "Microencapsulation of Living Cells and Tissues"; J. Pharm. Sci., 70, #4:351-354, April 1981.

11. Littlefield, S.G., Gilligan, K.J. and Jarvis, A.P.; "Growth and Monoclonal Antibody Production from Rat X Mouse Hybridomas: A Comparison of Microcapsule Culture with Conventional Suspension Culture": Damon Biotech Inc., Abstracts from the Third Annual Congress for Hybridoma Research, February 22, 1984, San Diego, CA, Hybridoma 3, #1:75, 1984.
12. Messing, R.A., Opperman, R.A. and Kolot, F.B.; "Pore Dimensions for Accumulating Biomass"; in Venkat, K., Immobilized Microbial Cells, pp. 13-26, ACS Symp. Series 106, 1979.
13. Nilsson, K., Scheirer, W., Merten, O.W., Ostberg, L., Liehl, E., Katinger, H.W.D. and Mosbach, K.; "Entrapment of Animal Cells for Production of Monoclonal Antibodies and Other Biochemicals"; Nature, 302:629-630, 1973.
14. Phillips, S.M. (University of Pennsylvania, School of Medicine, Philadelphia, PA); Private Communication; 1983.
15. Rupp, R.G.; "Large-scale Culture of Micro-encapsulated Cells"; Division of Microbial and Biochemical Technology, State of the Art Review Symposium on Large-Scale Mammalian Cell Culture, American Chemical Society Annual Meeting, Phildelphia, 1984.
16. Weinert, H. (Boston Biomedical Consultants, Waltham, MA); Private Communication; 1983.

PANEL DISCUSSION 2

DR. P. BROWN (Bio-Response, Inc., Hayward, CA): I have a question for Dr. Hopkinson, that deals with the growth of cells outside of 10,000 molecular weight cutoff fibers you described. I believe the implication there is that you're putting serum through the middle of those fibers and that those materials which pass through the fibers then support the growth of the cells. But our experience has been with 10,000 MW fibers, that no sort of animal derived nutrition can pass through to support cell growth unless the cells were loaded with residual serum.

DR. J. HOPKINSON (Amicon Corporation, Danvers, MA): What you asked is a very common question that we're asked, not only at 10,000 but also with a 50,000 cutoff. When cells are inoculated into the culture cartridge, they are inoculated into the extra capillary space outside the fibers in medium that contains serum. So at the beginning periods of the culture, you have complete serum surrounding the cells. With time, as you harvest off product, that serum is removed and is replaced only by serum ultrafiltrate. However, for several months thereafter, the serum levels and the initial concentrations drop and apparently those factors that are required at the early stages of the culture either are not required or are made in sufficient quantity later on that you do not need to supply complete serum to those cultures.

DR. G. BELFORT, (Rensselaer Polytechnic Institute, Troy, NY): I must disagree with you. We have run it on a 100K cutoff and a 10K cutoff and we have serious problems with a 10K cutoff growing an anti-mouse thymus surface antigen hybridoma line. We cannot get what we think are growth factors through the fibers to give it long-term performance. In fact, we've done that several times. We now have a 10K, 50K and 100K all going simultaneously, and I checked just when I left and it turned out that the 10K one was not healthy. There were problems with the cells. There could be other reasons, but the implications are that it may be growth factors that are not getting through. I want to ask you if you could tell us a little bit about the requirements and replacements of the medium, how frequently, what is an optimum replacement frequency of the medium for running these hollow fiber systems? We have done it every four days and we know that it may not be optimum to do it every two or every other day. I would like to get some comments on that.

DR. J. HOPKINSON (Amicon Corporation, Danvers, MA): Regarding your first comment, there have been a number of studies done comparing cell growth kinetics and output in different hollow fiber cutoffs. What you find in your particular case is not typical of what we find. I guess there are differences in different cell lines.

DR. G. BELFORT (Rensselaer Polytechnic Institute, Troy, NY): Have those been published, because I have not seen anything published anywhere.

DR. J. HOPKINSON (Amicon Corporation, Danvers, MA): No, there's nothing published at all. With regard to your second question, how frequently do you need to change medium, I guess you need to change it when you need to change it, but the question is determining what is the limiting factor, the toxic product that's building up and I think Bill addressed that. In our particular case, what we do is change medium every couple of days, sort of a complete change of the medium reservoir. We found that if you do it more frequently than that you do not gain anything with regard to the cell number. If you do it too much less frequently, you run into problems with pH and the cell growth isn't sufficient. But I don't think it's ever been looked at really from a scientific point of view of determining when you should change. Most people, when they first do these experiments, want to see if they can get the system to work, they're not interested in really optimizing it the first time around. So you want to make sure you're very lavish with your medium when you first get your feet wet with that technology and those methods.

DR. G. BELFORT (Rensselaer Polytechnic Institute, Troy, NY): You didn't comment in your talk on how to separate the antibody in the shell side from the cells. I would be very interested to know what do you suggest one does. There you have a high concentration of antibody being produced in a very thick cell concentrated suspension. How do you get the antibody out?

DR. J. HOPKINSON (Amicon Corporation, Danvers, MA): You're almost acting like my "straight man". Antibody is normally removed in the cartridge that I showed pictures of by ultrafiltering the extra capillary fluid out of the cartridge. This is simply done by opening the port in the extra capillary space and back pressuring the system. What this does is have a tendency to force fluid out of the extra capillary space and it is replaced by the ultrafiltrate. The rate at which that occurs is very, very slow, maybe in the range of 0.5-1.0 ml/min. That rate is not sufficient to force the cells out of the culture unit because they're so tightly packed in there. So that fluid just sort of weeps out of the system and you can turn over that extra capillary volume in about a half hour.

DR. G. BELFORT (Rensselaer Polytechnic Institute, Troy, NY): Just out of those ports? It just sort of comes out of the tubes and the cells don't move with it?

DR. J. HOPKINSON (Amicon Corporation, Danvers, MA): Very few cells come out at that time.

DR. C. HO (State Univ. of NY at Buffalo): I'd like to address this question to Dr. Hu. First, I'd like to compliment you on the excellent results. I've found that pH dictating the cell detachment is very intriguing. My question is, what kind of basic solution do you use in order to raise your pH?

DR. W. S. HU (University of Minneapolis, Minnesota): We use Hepes buffer. We have tried Hepes buffer and also PBS and just changed the pH essentially, because exposure time to trypsin is very low. So trypsin solution is made up in Hepes buffer at the proper pH.

DR. C. HO (State Univ. of NY at Buffalo): Now, if you change that basic solution, will the results change?

Dr. W. S. HU (University of Minneapolist, Minnesota): As I said, we have tested two kinds of buffer, either phosphate buffer or Hepes buffer. Both of those solutions work.

DR. J. FEDER (Monsanto Company, St. Louis, MO): I direct my comments to Dr. Dean. I think the table which you put together could be valuable. However, there is a problem in such a comparison in which different hybridomas were employed in the various systems studies. The monoclonal antibody production data could just as well reflect the differences in the hybridoma intrinsic expression capacities by the different systems used. It would have been valuable to have some reference production rates under a standard condition, such as in a T-flask or in suspension culture.

DR. R. C. DEAN, JR. (Verax Corporation, Hanover, NH): Well, what I tried to do was take your publications and I assumed that you are publishing tracked data. I also told you that for our system the actual measurements run from 0.5 micrograms/10^6 cells/hr (volume productivity) up to 2. So there's a range of 4-1. I also told you that we achieved as high as 800 micrograms/ml, but the average is more like 300. So for my system where I know what the data is, I gave you the ranges... for yours I don't know what the ranges are and people will have to interpret. All I can say is the stuff came out of the literature.

DR. J. FEDER (Monsanto Company, St. Louis, MO): Let me ask the question another way...if you took that same hybridoma in the simplest system, in a simple spinner at 10^6 cells/ml and you just let it set for 24 hours, what would your hybridoma produce in a 24-hour period per 10^6 cells?

DR. R. C. DEAN, JR. (Verax Corporation, Hanover, NH): I don't know. We don't run them in batch. I can tell you what they are in chemostats.

DR. J. FEDER (Monsanto Company, St. Louis, MO): Well, that's okay, chemostats maintained at what concentration?

DR. R. C. DEAN, JR. (Verax Corporation, Hanover, NH): I showed you a little bit of data there for a chemostat and by varying the dilution rate from .03 which was the top set of data down to .013 which is the bottom set of data, you may have noticed that the yield doubled. The cell concentration went up from something like 10^6 to 2.5 x 10^6 as the dilution rate went down. So there was a set of connected data. In that particular case, as I recall that curve of the yield 500 micrograms/ml was the upper point and the lower one was 250. We have seen 80 micrograms/ml and the highest are about 800 depending upon the operating conditions and the hybridoma.

DR. B. MAIORELLA (Cetus Corporation, Emoryville, CA): We have seen processes described today varying from using agitation with a traditional marine impeller through the 12 rpm sail and then things like the Amicon system with no agitation. How much trouble is there actually with sheer? How much of that is real and how much is mythology?

DR. W. R. TOLBERT (Monsanto Company, St. Louis, MO): Well, from my point of view, there is a significant problem with sheer. Going from my static system where there is no sheer and comparing to my agitated perfusion system where there is sheer, there is significantly more turnover of cells, more damage of cells even with a very gentle agitation system than there is without agitation. In the case of cells on microcarriers it's worse, because the cells are on the outside of the beads and experience whatever forces of collision right through the cells.

DR. R. C. DEAN, JR. (Verax Corporation, Hanover, NH): May I say something here? I don't know anything about sheer because we use fluidized bed systems which have pretty low sheer and try to protect our hybridomas inside our microcarriers. But I will tell you a little story. I used to work on blood pumps and ten years ago it was believed that turbulence destroyed red blood cells in pumping blood. Now it's known that that is not true. What destroys red blood cells is interaction with the walls of the vessels in which the red red blood cells are traveling. There are biochemical reactions that I'm not sure are thoroughly understood. But it has been proven pretty well that it's not the turbulence of the flow through the system or the sheer, but rather the interaction with the wall.

QUESTION: Dr. Dean mentioned projected cost in 1990 of $500/gram for antibodies. In view of some of the advances we've seen here today both in increased productivity and reduced medium cost, I wonder if you'd comment on whether that is correct or a little high. Can we do a little better than that?

DR. G. EDWARD (Celltech Ltd, Slough, Berkshire, United Kingdom): To get back to apples and oranges, unfortunately there's no easy way to actually compare all the different systems and the fairest way is one that we can't do, which is to actually say what's the cost of producing a gram of antibody. But we have already shipped antibody to major customers in the U.S. at below $500/gram. That's not our average price, that's been for large quantities of high producers, but those kinds of prices are certainly foreseeable in the not so distant future.

DR. W. R. TOLBERT (Monsanto Company, St. Louis, MO): I think we've seen in the last few years in the development of systems, but also in the development of serum free media and media supplements that it is not unreasonable to expect significant advances between now and 1990. It's pretty hard right now to give a measurement.

DR. J. HOPKINSON (Amicon Corporation, Danvers, MA): I'd just like to add one comment to that, too. There's a difference between the cost and the price. There are a number of companies here who sell monoclonal antibodies under contract and obviously their selling price is going to be higher than their costs. But I think there is a great deal of numbers that are tossed around the industry of the cost of monoclonal antibodies without backing that up with all the assumptions that go into that cost. For example, is that the cost for one gram or is that the cost of a kilogram, divided by a thousand. There are some substantial differences there. Certainly the numbers of several hundred dollars a gram up to a couple of thousand dollars a gram, depending on the output of the line is the sort of numbers that is tossed around right now. Certainly those numbers are going to drop as time goes on.

Index

N

O

P

R

S

T

V